T0367719

EVOLUTIONARY BIOLOGY

Evolution – both the fact that it occurred and the theory describing the mechanisms by which it occurred – is an intrinsic and central component in modern biology. Theodosius Dobzhansky captures this well in the much-quoted title of his 1973 paper, "Nothing in biology makes sense except in the light of evolution." The correctness of this assertion is even more obvious today: philosophers of biology and biologists agree that the fact of evolution is undeniable, and that the theory of evolution explains that fact. Such a theory has far-reaching implications. In this volume, twelve distinguished scholars address the conceptual, metaphysical, and epistemological richness of the theory and its ethical and religious impact, exploring topics including DNA barcoding, three grand challenges of human evolution, teleology, historicity, design, evolution and development, and religion and secular humanism. The volume will be of great interest to those studying philosophy of biology and evolutionary biology.

R. PAUL THOMPSON is Professor in the Institute for the History and Philosophy of Science and Technology and the Department of Ecology and Evolutionary Biology at the University of Toronto. His most recent books include *The Structure of Biological Theories* (1989) and *Agro-Technology* (Cambridge, 2011), and he is editor of *Issues in Evolutionary Ethics* (1995).

DENIS WALSH holds the Canada Research Chair in the Philosophy of Biology and is a member of the Department of Philosophy, the Institute for the History and Philosophy of Science and Technology, and the Department for Ecology and Evolutionary Biology at the University of Toronto. He is editor of *Naturalism, Evolution and Mind* (Cambridge, 2001).

EVOLUTIONARY BIOLOGY

Conceptual, Ethical, and Religious Issues

EDITED BY

R. PAUL THOMPSON

University of Toronto

and

DENIS WALSH

University of Toronto

CAMBRIDGE
UNIVERSITY PRESS

University Printing House, Cambridge CB2 8BS, United Kingdom

Cambridge University Press is part of the University of Cambridge.

It furthers the University's mission by disseminating knowledge in the pursuit of education, learning and research at the highest international levels of excellence.

www.cambridge.org
Information on this title: www.cambridge.org/9781107027015

© Cambridge University Press 2014

This publication is in copyright. Subject to statutory exception and to the provisions of relevant collective licensing agreements, no reproduction of any part may take place without the written permission of Cambridge University Press.

First published 2014

A catalogue record for this publication is available from the British Library

Library of Congress Cataloguing in Publication data
Evolutionary biology : conceptual, ethical, and religious issues / edited by R. Paul Thompson and Denis Walsh.
pages cm
Includes bibliographical references and index.
ISBN 978-1-107-02701-5 (hardback)
1. Biology–Philosophy. 2. Evolution (Biology)   I. Thompson, R. Paul, 1947–, editor of compilation.
QH331.E87   2014
576.8–dc23   2013039683

ISBN 978-1-107-02701-5 Hardback

Cambridge University Press has no responsibility for the persistence or accuracy of URLs for external or third-party internet websites referred to in this publication, and does not guarantee that any content on such websites is, or will remain, accurate or appropriate.

For Michael Edward Ruse
Intellectual pioneer, a founder of modern philosophy of biology,
dedicated student mentor and a warm supportive friend to many

Contents

List of contributors   *page* ix
Acknowledgments   xi

Introduction   1

1   Human evolution: whence and whither?   13
     Francisco Ayala

PART I   EVOLUTION AND THEOLOGY   29

2   Evolutionary theory, causal completeness, and theism:
     the case of "guided" mutation   31
     Elliott Sober

3   Religion, truth, and progress   45
     Philip Kitcher

PART II   TAXONOMY AND SYSTEMATICS   63

4   Consilience, historicity, and the species problem   65
     Marc Ereshefsky

5   DNA barcoding and taxonomic practice   87
     David Castle

PART III   THE STRUCTURE OF EVOLUTIONARY THEORY   107

6   Darwin's theory and the value of mathematical
     formalization   109
     R. Paul Thompson

7   Population genetics, economic theory, and eugenics in
    R. A. Fisher   137
    Jean Gayon

8   Exploring development and evolution on the tangled bank   151
    Jane Maienschein and Manfred Laubichler

PART IV FUNCTION, ADAPTATION, AND DESIGN   173

9   Darwin's cyclopean architect   175
    John Beatty

10 Function and teleology   193
    Denis Walsh

11 How physics fakes design   217
    Alex Rosenberg

Index   239

Contributors

FRANCISCO AYALA is University Professor and Donald Bren Professor in the Department of Ecology and Evolutionary Biology at the University of California, Irvine.

JOHN BEATTY is Professor in the Department of Philosophy at the University of British Columbia.

DAVID CASTLE is Professor at the ESRC Innogen Centre at the University of Edinburgh.

MARC ERESHEFSKY is Professor in the Department of Philosophy at the University of Calgary.

JEAN GAYON is Professor at the Institut d'histoire et de philosophie des sciences et des techniques at the Université Paris 1 Panthéon-Sorbonne.

PHILIP KITCHER is John Dewey Professor of Philosophy at Columbia University.

MANFRED LAUBICHLER is Professor at the Center for Biology and Society, and Center for Social Dynamics and Complexity at Arizona State University.

JANE MAIENSCHEIN is Professor, Regents' Professor, and President's Professor at the Center for Biology and Society and School of Life Sciences at Arizona State University.

ALEX ROSENBERG is Chair and R. Taylor Cole Professor of Philosophy at the Department of Philosophy, Duke University.

ELLIOTT SOBER is Professor, Hans Reichenbach Professor, and William F. Vilas Research Professor at the Department of Philosophy at the University of Wisconsin–Madison.

R. PAUL THOMPSON is Professor at the Institute for the History and Philosophy of Science and Technology, the Department of Ecology and Evolutionary Biology, and the Department of Philosophy, at the University of Toronto.

DENIS WALSH is Professor at the Institute for the History and Philosophy of Science and Technology, Department of Philosophy, and the Department of Ecology and Evolutionary Biology, at the University of Toronto.

Acknowledgments

Evolutionary theory has a complex and fascinating history, and it is conceptually and theoretically rich. Hence, it is not surprising that historians, philosophers, and biologists have mined the rich veins of gold it contains. As this volume demonstrates there is considerable gold left to be unearthed.

There is a worthy tradition of thanking those who, directly or indirectly, have had an impact on a volume. Our list is too extensive to make specific mention practicable. Those that have contributed chapters have, obviously, had a crucial impact on its quality and relevance. Those who provided helpful comments on the proposal and the text have improved the final result. We would like to acknowledge the serious health issues that prevented Elisabeth Lloyd and Robert Brandon from submitting chapters. The staff of Cambridge University Press have made important contributions to the accuracy, readability, and style of the volume: Hilary Gaskin (editor), Emma Walker and Anna Lowe, in particular. Also, many thanks to Sylvia Nickerson for the artwork in Chapter 9, Fermin Fulda for compiling the index, and Alison Evans of Out of House Publishing. Those who know our spouses, Jennifer McShane and Deborah Kohn, are familiar with their constant support and encouragement; for others we acknowledge here their support and endurance. Although having no specific hand in this volume, always lurking in the background is the indefatigable Michael Ruse.

Introduction

Contemporary analytic philosophy of biology was forged in the 1960s. It began a little more than 50 years ago with Morton Beckner's *The Biological Way of Thought* (1959). Building on this seminal contribution, in articles and books, Thomas Goudge (*The Ascent of Life*, 1961), Marjorie Green (*Approaches to a Philosophical Biology*, 1968), David Hull (*Philosophy of Biological Science*, 1974), and Michael Ruse (*The Philosophy of Biology*, 1973) laid the foundation for modern philosophy of biology.[1] These founders of the field articulated and staked out positions on nearly all the important logical and conceptual underpinnings of evolutionary biology, as well as the social implications of its theories and empirical discoveries.

Michael Ruse's 1973 *Philosophy of Biology* consolidated the field by providing a rigorous analysis and comprehensive treatment of nearly all the critical conceptual issues, including those that have remained contentious; it still stands as a *tour de force*. In 1979, *The Darwinian Revolution: Science Red in Tooth and Claw* was published. It remains an exemplar of the integration of philosophy of science and history of science. Since that time, he has:

- founded, in 1986, the leading journal in philosophy of biology, *Biology and Philosophy* (and nurtured it into being one of the top four journals in philosophy of science);
- founded, in 1995, and edited, from 1995 to 2011, the *Cambridge Studies in Philosophy and Biology* series, which during that period published 80 of the most important books in the field;

---

[1] A few biologists – J. H. Woodger, C. H. Waddington, and Bernhard Rentch, for example – and physicists – Erwin Schrödinger, for instance – had tackled philosophical aspects of biology but philosophical interest in biology by philosophers of science dates from the work of this group. Earlier philosophical work such as Henri Bergson's *Creative Evolution* and the use by philosophers of Darwinian fitness and Lamarckian inheritance, such as by Herbert Spencer, are very different from contemporary analytic philosophy of biology.

- written more than 20 books (almost all of which have been translated into other languages);
- edited more than a dozen books;
- contributed more than 100 journal articles;
- been a leader in championing evolution in the broader society and in promoting science education.

Moreover, his impact on philosophy of biology includes mentoring several generations of researchers and scholars who have achieved international reputations in their own right. He has received numerous prestigious research awards, including the John Simon Guggenheim Fellowship and Isaak Walton Killam Fellowship. He was elected Fellow of the Royal Society of Canada and Fellow of the American Association for the Advancement of Science, and has received honorary degrees from the University of Bergen, McMaster University, and the University of New Brunswick.

Given his formative role in the development of philosophy of biology, his contributions to research and scholarship, his broader social contributions, his mentoring of generations of scholars and researchers, and his impressive publication record and influence, it is fitting that this volume of original articles by internationally renowned philosophers of biology should be dedicated to him. Although some of the contributors to this volume disagree with some of his positions and arguments, all recognize his importance and the profound impact he has had on the field; many make direct reference to his work. As Michael has told so many of us over the last 50-plus years, "criticize me; just don't ignore me." He has certainly not been ignored and there is no shortage of criticism.

This volume continues the exploration of evolutionary biology that he initiated. Today evolution – both the fact that it occurred and the theory, descended from Darwin, describing the mechanisms by which it occurred – is an intrinsic and central component in modern biology. Theodosius Dobzhansky captures this well in the oft-quoted title of one of his 1973 papers,[2] "Nothing in biology makes sense except in the light of evolution." The correctness of this assertion is even more obvious today than in 1973. Philosophers of biology, historians of biology, and biologists agree that the fact of evolution is undeniable, and that the theory of evolution provides unity to evolutionary biology as a whole, is conceptually rich, and has far-reaching social implications. Like all scientific theories,

---

[2] Dobzhansky 1973.

however, there are some conceptual and epistemological underpinnings on which there is no settled opinion. Also, like all sciences, there are implications of evolutionary biology that engender intense public controversy.

Notwithstanding the central place of evolutionary theory in biology, there are a number of conceptual and epistemological underpinnings on which there is no settled opinion. These include: the relationship of organisms and their molecular components, the nature of species, the nature of adaptation, the formal (logical/mathematical) structure of evolutionary theory, and the nature and role of development. Each of these poses deep philosophical challenges. The chapters in this volume continue and advance the discussion of them.

The contributors to this volume are philosophers and biologists who have been at the forefront of seeking resolutions to these pivotal conceptual and societal issues. With the exception of the tension between evolution and certain religious sects, there has been considerable convergence, over the last 50 years, with respect to all these issues. Sometimes the convergence has moved debate closer to resolution; sometimes it has led to an identification of remaining impediments. In the case of the tension between evolution and literalist fundamentalist Christianity and Islam, the nature of the tensions and the critical importance of resolving them have been brought into sharper focus. The goal of the volume is to provide readers with a window on the current thinking of those who have shaped the discourse on these contentious issues over several decades.

The collection begins with a contribution from the eminent evolutionary biologist Francisco Ayala. Professor Ayala has a longstanding history of collaboration with Michael Ruse, and his chapter demonstrates the rich potential to be found in the cross-pollination between philosophy and evolutionary biology that Ruse has done so much to foster. Ayala takes up themes broached in Ruse's most recent book, *The Philosophy of Human Evolution* (2012). Specifically, Ayala addresses the evolution of ethical behavior in the transition from ape to human. Ethical behavior has clearly evolved, but quite how it might have done so has been a challenge to evolutionists. There are two principal problems for any evolutionary ethics. The first is that the standard strategy deployed in explaining the evolution of some structure or ability appears to break down in the case of the human capacity for moral judgment and action. Typically, to explain the conditions under which some feature has evolved, one simply articulates the fitness benefit that feature confers on its bearers. The vexed problem for evolutionary ethics is that moral imperatives and fitness imperatives don't obviously coincide. The second problem is what Ayala calls the "naturalistic fallacy." Those who seek to ground ethical behavior in evolution run the risk of negating it. If ethical behavior consists in acting ultimately on fitness imperatives, then we have merely been duped by our genes into thinking we are acting under the guise of the moral good.

Ayala's chapter seeks to finesse these two problems simultaneously. He distinguishes between two questions that are often conflated: (1) whether our capacity for moral deliberation and behavior is an evolutionary endowment, and (2) whether the specific moral norms that guide our actions are an evolutionary endowment. Ayala delivers a positive verdict on the first question: "Humans evaluate their behavior as either right or wrong, moral or immoral, as a consequence of their eminent intellectual capacities, which include self-awareness and abstract thinking. These intellectual capacities are products of the evolutionary process, but they are distinctively human" (p. 18). But, in opposition to much of sociobiology and mainstream evolutionary ethics, he insists upon a negative answer to the second: "moral norms according to which we evaluate particular actions as morally either good or bad ... are products of cultural evolution, not of biological evolution. The norms of morality belong, in this respect, to the same category of phenomena as the languages spoken by different peoples, their political and religious institutions, and the arts, sciences, and technology" (p. 18).

The capacity for ethical behavior, Ayala argues, is conferred on us by three distinctively human cognitive abilities: the ability to anticipate consequences, the ability to make value judgments, and the ability to choose between available courses of action. While these abilities are jointly constitutive of the capacity for ethical behavior, they are not exclusively moral faculties. They grow out of the facility that our hominin ancestors developed for the use and production of tools, means–end reasoning, the planning and assessment of other forms of action. Ayala sees "no evidence that ethical behavior developed because it was adaptive in itself ... It seems rather that the likely target of natural selection was the development of advanced intellectual capacities" (p. 22).

After Francisco Ayala's tour through the challenges facing the study of human evolution, Part I of this collection turns to an area of dispute in which Michael Ruse has become particularly prominent in recent years: the compatibility of evolutionary biology with religious thought. Ruse has been perhaps the pre-eminent exponent of conciliation between the power of evolutionary biology to reveal the mysteries of life, and the draw many feel toward devotional religious belief. Ruse has consistently valued irenics over histrionics on these matters; his has been the voice of moderation heard over the stentorian tones emanating from both secular and religious extremes. His *Evolution and Religion: A Dialogue* (2008) captures all these tones with a keen composer's ear. This section features two eminent philosophers of science, Elliott Sober and Philip Kitcher. Each in his way comes out strongly in support of both Ruse's placatory tone, and his compatibilist message.

Elliott Sober argues that evolutionary theory is logically consistent with the conception of a god familiar to Abrahamic religions, who intervenes in the processes of the world. Sober explains that evolutionary theory is fundamentally probabilistic. The theory yields probabilities of certain outcomes – for instance the increase of one trait type over another – given certain conditions. As Sober points out, probabilistic theories can be true, inductively generalizable, testable, and informative, even if they are not causally complete. There is room then, for "hidden variables," causes of evolutionary phenomena that are not articulated by the theory. It is *logically* consistent, then, with evolutionary theory that these unarticulated causes may be supernatural in origin. Nothing in evolutionary theory proscribes this. It is commonly thought that if there is divine involvement in the processes of evolution, it should be manifested in the pattern of evolutionary novelties. A providential god should or would bring about novelties that are beneficial to those organisms in which they arise. Biologists know, however, that evolutionary novelties arise through mutations, and that mutations are random – in the sense of unguided. But the unguidedness of mutations is in no way incompatible with the thesis that they are divinely caused: "[W]hat biologists mean, or ought to mean, when they say that mutations are unguided says nothing about whether God ever causes a mutation to occur" (p. 32). Invoking Pierre Duhem, Sober reminds us that the application of a scientific theory to the world requires auxiliary assumptions. Evolutionary theory could only have implications about the existence of a deity if it were supplemented by certain auxiliary assumptions. But these auxiliary assumptions are all philosophical, and not biological, in nature. They are not licensed by evolutionary theory alone. Striking a note strongly concordant with Ruse's own message, Sober concludes: "Atheists who think that evolutionary theory provides the beginning of an argument for disbelieving in God should make clear that their arguments depend on additional premises that are not vouchsafed by scientific theory or data" (p. 43).

Philip Kitcher in his chapter addresses the delicate issue of reconciling the role of religion with the atheist's conviction that religious beliefs are false. Kitcher aligns himself with Ruse here, against a phalanx of outspoken contemporary atheists, particularly the self-anointed "four Horsemen." "Although we agree with the Horsemen that there is a sense in which all religious doctrines are false, we don't take this to be the end of the proper discussion of religion" (p. 46). Kitcher draws upon the pragmatist insight that the world we inhabit is to a significant degree one of our own making, structured by "our psychological faculties and our purposes" (p. 48). As scientists we pursue the process of comprehending, predicting, and intervening on the world. And we have generated ways of thinking appropriate to those purposes. "But these are not our only purposes," Kitcher reminds us (p. 49). "We devise ways of thinking and forms of language directed toward different ends – in play, in literature and arts, in ethics, and in religion" (p. 49). Kitcher introduces us to the idea that truth in general applies to sentences that are employed in the pursuit of a human project. As such there exist a range of species of truth. Kitcher outlines his conception of religious truth. *S* is weakly religiously true (roughly) just when there is an established religious practice that affirms *S*. *S* is strongly religiously true (again, roughly) just when any progressive modification of said practice would continue to affirm *S*. There are weakly religious truths. Kitcher speculates that there may be no strongly religious truths (except, perhaps, for strong ethical truths). He doubts "whether any particular fiction, even the myths of the axial age, is so deep and fundamental that it delivers strong religious truths" (p. 60). That said, religious practice will rightly continue to form a part of the human project of responding to the challenges of "forging identities" and "achieving communities". These projects are in no way incompatible with the scientific project of comprehending, predicting and intervening on the world.

Part II focuses on taxonomy and systematics, another topic on which Ruse has made many important contributions. The concept "species" is central to taxonomy and has been a thorny concept since before the publication of the *Origin*. Darwin spends much of the first three chapters of the *Origin* arguing for, essentially, a nominalist conception; that is, species are not real, they are a human artifact that is useful – perhaps essential – to biology but not part of the nature of things. Contemporary evolutionary biologists recognize a number of different – and not necessarily compatible – definitions of "species." The most commonly known is the biological species concept; members of the same species can interbreed without sterility. This has proved a useful definition in a number of contexts but does not apply to non-sexually reproducing organisms, and they comprise most of the living world. Moreover, it doesn't even apply in many cases of sexually reproducing organisms. The classic exception is ring species and there are numerous instances of ring species.

Marc Ereshefsky responds directly to Michael Ruse's work on the species concept. He identifies two philosophical issues that Ruse has addressed. One is the ontological status of species: are species natural kinds akin to elements on the periodic table or are species individuals akin to particular organisms? The other concerns whether "species" refers to a real category in nature or whether the species category is merely an artifact of our theorizing. On both issues, he contends that Ruse made major and important contributions. Nonetheless, although Ruse's arguments concerning species are cogent and innovative, Ereshefsky contends that they are flawed. He mounts a case for considering species as historical entities, something to which, he contends, Ruse pays too little attention. On the question whether "species" refers to a real category in nature, he offers a pragmatic form of species anti-realism.

David Castle explores the nature and role of DNA barcoding in taxonomic practice. Barcoding is relatively new in taxonomy. As the term suggests, DNA barcoding is similar to merchandise barcoding except the "bars" are short segments of DNA rather than lines of different lengths, thicknesses, and spacing. Hence, barcoding provides a method by which groups of organisms can be differentiated by comparing short, standardized regions of DNA. Castle examines the how the taxonomic community has responded to DNA barcoding; to state that this technique is still controversial is to understate the polarization it has created. A pluralist perspective seems appropriate – that is, traditional taxonomic practice and barcoding informing each other – but that has yet to be achieved. Castle opens his chapter with a very useful introduction to barcoding, its aims and methods. He then examines three main objections to barcoding. His position centers on barcoding as an evolving method and he sees the objections to it as, in significant part, being motivated by protection of past practices; barcoding, "exemplifies to traditional taxonomists many perceived threats they most fear." The outcome of his analysis should lay the groundwork for a pluralistic approach and allay the fear of traditional taxonomists.

Part III focuses on the structure of evolutionary theory. Two views on the structure of evolutionary theory dominated philosophical discussion in the 1960s. One, strongly influenced by logical empiricism, maintained that the logical structure of theories in biology was the same as physics; both sought axiomatic-deductive systems of laws, which explained and predicted phenomena by deducing them using the laws of the theory and relevant conditions. This is the view Ruse championed in his 1973 book. He provided a sketch of the axiomatic and deductive nature of evolutionary theory and gave examples of explanatory deductions. The other view maintained that evolutionary theory was different from theories in physics. The nature of those differences varied. Thomas Goudge's "narrative explanation" is an example of a non-deductive explanatory pattern. Ruse contended that narrative explanations were enthymemes and when complete were in fact deductive. Paul Thompson and Jean Gayon essentially agree with Ruse that theories and explanations in biology and in physics are for the most part the same. They, however, are not committed to a logical empiricist conception of either physics or biology. They also recognize that in physics and biology there are numerous metaphors and analogies that motivate and interpret concepts, as Jean Gayon's chapter underscores. And they hold that mathematics provides the language and structure for theories, as Paul Thompson's chapter emphasizes. Thompson's and Gayon's chapters are focused on the genetical theory of evolution, but, in the last 30 years, it has become apparent that a complete causal account of evolution must embrace development. Just how that should be done and what the resulting causal structure of evolutionary theory will look like is emerging, as Jane Maienschein and Manfred Laubichler's chapter makes clear.

In his chapter, Thompson develops a response to an obvious, and frequently voiced, criticism regarding his claim that mathematics is the language of scientific theories. In *The Origin of Species*, Darwin employs no mathematics, and yet he formulated a theory that is central to modern biology; this is beyond doubt. Thompson claims that Darwin provided a brilliant "informal" theory along with a wealth of evidence. It was not, however, until a mathematical "formal" account was given in the late 1920s that the internal structure of the theory and its empirical implications were clearly understood. The debates, in the 50 years after the publication of *The Origin*, about whether selection acting on small individual variations could lead to evolutionary change, whether selection decreased variation, whether selection and Mendelian heredity are compatible – or not, as Bateson claimed – and the like were only resolved when a mathematical formalization was provided that integrated Darwinian theory, Mendel's theory, and biometry into a single theory.

In his chapter, Gayon explores R. A. Fisher's analogical use of economic theory in both his population genetics work and his eugenics. Gayon provides numerous examples and weaves a compelling argument that economic analogies connect these two strands of his work and, moreover, that his eugenic ideas motivated the way he defined and explicated natural selection in his *Genetical Theory of Natural Selection*. Gayon examines in detail the analogy between growth of a population (Fisher's Malthusian parameter) and growth of capital, thereby demonstrating that "Fisher's demographic approach to natural selection relied upon an explicit economic analogy: making children is interpreted in terms of investment, cost, benefit, and repayment." The same dependence on an economic analogy is found in his eugenics writings. The core insight of this chapter is that "Fisher's economic interpretation of the Malthusian parameter was motivated, or at least inspired, not by abstract considerations about natural selection, but by his eugenic way of thinking" (p. 143).

Jane Maienschein and Manfred Laubichler discuss the philosophical and theoretical implications of one of the most exciting and expansive areas of biology to emerge in the last quarter-century – evolutionary developmental biology. Rightly or wrongly, Modern Synthesis lore has it that organismal development was left out of the synthesis. This may not be wholly historically accurate, but it seems fair to say that a proper understanding of the importance of development for evolution has only recently emerged. But quite where this significant expansion of our understanding leaves evolutionary theory is a matter of some debate. Some authors see the assimilation of development into evolution (evo-devo) as wholly complementary to accepted evolutionary theory, as a completion of the project of synthesizing biology's various sub-disciplines embarked on so boldly in the 1930s. "Within evo-devo, the logical place of development within evolutionary theory was in explaining the details of the genotype-phenotype map without changing the explanatory structure of evolutionary biology, which, at its core, was still based on population dynamics" (p. 162). But there is another interpretation; developmental evolutionary biology ('devo-evo') sees tensions. Developmental evolutionary biology promotes a more radical reorientation of evolutionary theory: "[I]n the context of developmental evolution the causal structure of evolutionary explanation has shifted from a primacy of population-level dynamics to the primacy of developmental mechanisms and that explaining the origin of variation rather than the fate of variants within populations is the first and most important problem for all theories of phenotypic evolution" (p. 168).

Adaptation, teleology, and design were at the heart of the *Origin* and are still contentious concepts in evolutionary theory. Part IV explores why living things are the only non-artifacts about which talk of design and purpose seems appropriate – even perhaps indispensable. Organisms are adapted to their conditions of existence – sometimes extravagantly so – and this alone sets them apart from the rest of the natural world. Indeed reconciling the place of organisms as products of blind, natural processes with their evident exquisite adaptations is a central challenge to evolutionary biology and its philosophy. It was a central objective of Darwin's theory, and it continues to challenge and perplex. Evolution, we are told, is fundamentally chancy, and yet the very concepts of *design and purpose* are in diametrical opposition. In this section three philosophers of biology explore the chance/purpose tension.

John Beatty discusses the metaphor of selection as an architect, found repeatedly throughout Darwin's writing. There are two components to building a construction: materials and a design. Darwin likens variation to the former and natural selection to the latter. But the analogy is multiple and ambiguous and, as beautifully documented by Beatty, is used in a variety of ways by Darwin himself. The ambiguities point to a long-standing question in both the interpretation of Darwin and in modern evolutionary biology concerning the relative importance of variation over selection in the explanation of form. On a strong – nowadays called 'adaptationist' – reading the sources of variation may be explanatorily negligible in comparison to selection. Beatty says: "If the causes of variation do not bias it toward (or away) from the direction of selection, then the direction of evolution by natural selection would seem to be unaided (and unimpeded) by the course of variation, and thus natural selection would seem to be solely responsible for the outcome" (p. 179). But the architect analogy does not support the primacy of selection over the sources of variation univocally. "Whether ... the architect analogy demonstrates the major importance of natural selection, and the minor importance of the production of variation, depends on how the analogy is interpreted" (p. 188). Like Darwin's theory itself, nothing about his recurrent architect metaphor commends the primacy of natural selection over chance variation in the explanation of adaptive form.

Denis Walsh discusses the alleged reduction of teleological concepts like function and purpose in evolutionary biology. Evolutionary biology, as Michel Ruse has often pointed out, is peppered with teleological-sounding talk of functions and purposes. Most commentators believe that the status of evolutionary biology as a science depends upon explaining this teleology away – ersatz teleology is acceptable, real teleology is taboo. The standard way of de-teleologizing biological talk is to interpret all teleological locutions as instances of a particular kind of historical explanation that adverts to the effects of natural selection in the past. This is the Etiological Theory of Function, first introduced and promoted by Michael

Ruse himself. Having arisen in the 1970s, the etiological theory has had a long run, and is now the accepted method for sanitizing biological teleology. Walsh argues that the etiological theory displays the hallmarks of both the philosophy and the biology of the time of its inception. It betrays its philosophical provenance by the supposition that what is needed is a conceptual analysis – a translation of all teleological talk. It also harks back to a biology that predates the newly emerging appreciation of the roles played by organismal processes in the production of evolutionary novelties, in securing the high fidelity of inheritance, and even in maintaining the structural integrity of the genome. Freed from these outdated constraints, Walsh believes that there is no need for a non-teleological analysis of teleological talk. Organisms are fundamentally purposive entities, and an understanding of this is crucial to an understanding of the process of adaptive evolution.

Alexander Rosenberg takes a significantly different tack. Long ago Erwin Schrödinger (1944) pointed out what appears to be a paradox of life. Living things appear to violate the second law of thermodynamics. That law tells us that closed systems inexorably move toward a state of maximum disorder. But organisms are highly ordered, self-building, radically negentropic systems. Adaptation and design appear to violate the basic laws of physics. Rosenberg takes up the theme. In doing so, he is laying down a challenge to the anti-reductionist consensus in the philosophy of biology. The Unity of Science hypothesis (Oppenheim and Putnam 1958) avers that all phenomena can be explained by physics alone. Rosenberg's account of adaptation offers a vivid example: "Physics is sufficient for evolution by natural selection … [Starting with] the second law we can show that the process Darwin discovered is, necessarily, the only way adaptations can emerge, persist and be enhanced in a world where the physical facts fix all the facts" (p. 219). Organisms may be negentropic, but adaptive evolution is a splendidly efficient way to increase entropy. "It's hard to think of a better way to waste energy than to produce lots of energetically expensive copies of something and then destroy all of them except for the minimum number of copies that you need to do it all over again" (pp. 229). Indeed, Rosenberg avers that the only way to harness the second law in building, propagating, and enhancing adaptations is through the process that Darwin discusses. Rosenberg ends with a much less concessive gesture toward theism than is found in either the Sober or Kitcher offerings. According to Rosenberg, a proper understanding of the physics of adaptation shows that evolutionary theory is deeply incompatible with the theism of the Abrahamic religions.

REFERENCES

Beckner, Morton (1959) *The Biological Way of Thought*. New York: Columbia University Press.

Dobzhansky, Theodosius (1973) "Nothing in Biology Makes Sense Except in the Light of Evolution." *American Biology Teacher* 35: 125–29.

Goudge, T. A. (1961) *The Ascent of Life: A Philosophical Study of the Theory of Evolution*. University of Toronto Press.

Green, Marjorie (1968) *Approaches to a Philosophical Biology*. New York: Basic Books.

Hull, David (1974) *Philosophy of Biological Science*. Englewood Cliffs, NJ: Prentice-Hall.

Oppenheim, P. and H. Putnam (1958) "The Unity of Science as a Working Hypothesis." In H. Feigl, Michael Scriven, and Grover Maxwell (eds.), *Minnesota Studies in the Philosophy of Science*, vol. II. Minneapolis: Minnesota University Press, pp. 3–36.

Ruse, Michael (1973) *The Philosophy of Biology*. London: Hutchinson.

    (2008) *Evolution and Religion: A Dialogue*. Lanham, MD: Rowman & Littlefield.

    (2012) *The Philosophy of Human Evolution*. Cambridge University Press.

Schrödinger, Erwin (1944) *What is Life? The Physical Aspect of the Living Cell*. Cambridge University Press.

CHAPTER I

Human evolution
Whence and whither?

Francisco Ayala

This chapter is dedicated to Michael Ruse and refers to his 2012 book, *The Philosophy of Human Evolution*, among the most recent in his large and distinguished bibliography, counting several dozen books concerning the philosophy of biology, broadly understood.

I first knew about Michael Ruse in 1973, when I found his *The Philosophy of Biology*, published that year. In chapter 9, Ruse refers to and quotes from a discussion of functional statements in my 1968 paper, "Biology as an Autonomous Science," but he does not further refer to my extended treatments of teleological explanations in biology and of the reduction of theories in the same paper or some other of my early publications. He would later often refer to my philosophical publications, although I suspect always with the reservation that I am a biologist, not a *bona fide* member of the philosophical profession. He has always been generous, nevertheless. In the bibliography to his 1988 *Philosophy of Biology Today*, he listed 29 works of mine (more than those of any author, other than David Hull and Ruse himself), including the book that I edited in 1974 with Theodosius Dobzhansky, *Studies in the Philosophy of Biology*, written before I had the good fortune of meeting Michael. He has continued over the years to be generous in praise and otherwise, inviting me in numerous instances to contribute to books that he has edited over the years. He has contributed also to books I have edited, notably recently, "The Darwinian Revolution: Rethinking Its Meaning and Significance," in *In the Light of Evolution*, vol. III, *Two Centuries of Darwin* (Avise and Ayala 2009) and "The Biological Sciences Can Act as a Ground for Ethics," in *Contemporary Debates in Philosophy of Biology* (Ayala and Arp 2010).

I have read, over the years, many articles and books written by Michael Ruse and learned much from them. Many of his books, 34 of them, fill a full shelf in my library. Most of them, I am pleased to acknowledge, have a hand-written dedication, "with warm friendship" and the like. Most of them also have my own annotations and page references. In 1997,

I published in *Science* a review of his magnificent *Monad to Man: The Concept of Progress in Evolutionary Biology* (1996). Recently, I have been most pleased because Ruse has formally dedicated to me his most recent (as of this writing) book, *The Philosophy of Human Evolution* (2012). It is because of this that I have decided to write my chapter for this book on that subject.

In the dedicatory in *The Philosophy of Human Evolution*, Michael Ruse refers to our friendship that he has enjoyed "over thirty years." It is a friendship I have much enjoyed as well, while also admiring his immense productivity and distinguished scholarship, which place him, as I am sure most experts would agree, at the very top, in the United States and in the world, of the discipline of philosophy of biology. It is from that enduring friendship and admiration that I dedicate this chapter to Michael Ruse.

Introduction

Mankind is a biological species that has evolved from other species that were not human. In order to understand human nature, we must know our biological make-up and whence we come, the story of our humble beginnings. For a century after the publication of Darwin's *On the Origin of Species* in 1859, the story of evolution was reconstructed with evidence from paleontology (the study of fossils), biogeography (the study of the geographical distribution of organisms), and from the comparative study of living organisms: their morphology, development, physiology, and the like. Since the mid twentieth century we have had, in addition, molecular biology – the most informative and precise discipline for reconstructing the ancestral relationships of living species. Molecular biology has shown that our closest relatives are the chimpanzees, who are more related to us, and we to them, than they are to gorillas and much more than to orangutans.

The deciphering of the human and chimpanzee genomes and other molecular biology exploits have shone much light on what it is that makes us humans and how we got here through the evolution of our genetic make-up and how our genome differs from the genome of our closest relatives, the chimpanzees. Surely, the continuing analysis of the genome of human and other animals will throw much further light on understanding human nature.

Neandertal hominins (*Homo neanderthalensis*), with brains as large as those of *H. sapiens*, appeared in Europe around 400,000 years ago (kya) and persisted until 40 kya. The Neandertals were thought to be ancestral to anatomically modern humans, but now we know that modern humans appeared at least 100 kya, much before the disappearance of the Neandertals. Moreover, in caves in the Middle East, fossils of modern humans have been found dated at nearly 100 kya, as well as Neandertals dated at 60 and 70 kya, followed again by modern humans dated at 40 kya. It is unclear whether the two forms repeatedly replaced each other by migration from other regions, or whether they coexisted in the same areas.

Molecular biology has recently provided genetic evidence indicating that some interbreeding between *H. sapiens and H. neanderthalensis* likely occurred. Approximately, 2.5 percent of the genome of non-African modern humans derives from the Neandertal genome (Green *et al.* 2010; Stringer 2012).

One more human species has come to play in recent years, the Denisovans. DNA sequencing of a tiny sample from a fossil finger found in the Denisova cave in Siberia has identified this new species, more closely related to the Neandertals than to modern humans (Reich *et al.* 2010; Meyer *et al.* 2012). Surprising also is that about somewhat more than 2 percent of the DNA of modern humans from Australia and New Guinea derives from Denisovan DNA (Reich *et al.* 2010; Rasmussen *et al.* 2011; Skoglund and Jakobsson 2011; Meyer *et al.* 2012). Yet one more unanticipated outcome, obtained by "high-throughput" DNA sequencing, is the discovery that between 2 and 5 percent of the genome of modern humans from sub-Saharan African populations derives from older hominin populations, as yet unidentified as to their species (Hammer *et al.* 2011).

Humankind's distinctive traits

Erect posture and large brain are the two most conspicuous human anatomical traits. We are the only vertebrate species with a bipedal gait and erect posture; birds are bipedal, but their backbone stands horizontal rather than vertical (penguins are a minor exception). Brain size is generally proportional to body size; relative to body mass, humans have the largest (and most complex) brain. The chimpanzee's brain weighs less than a pound; a gorilla's slightly more. The human male adult brain has a volume of 1,400 cubic centimeters (cc), about three pounds in weight.

Until recently, evolutionists raised the question whether bipedal gait or large brain came first, or whether they evolved consonantly. The issue is now resolved. Our *Australopithecus* ancestors had, since 4 million years ago, a bipedal gait, but a small brain, about 450 cc, a pound in weight.

Brain size starts to increase notably with our *Homo habilis* ancestors, about 2.5 million years ago, who had a brain about 650 cc and also were prolific tool-makers (hence the name *habilis*). Between 1 and 2 million years afterwards, *Homo erectus* had adult brains of about 1,200 cc. Our species, *Homo sapiens*, has a brain about three times as large as that of *Australopithecus*, 1,300–1,400 cc, or some three pounds of gray matter. Our brain is not only much larger than that of chimpanzees or gorillas, but also much more complex. The cerebral cortex, where the higher cognitive functions are processed, is in humans disproportionally much greater than the rest of the brain when compared to apes.

Erect posture and large brain are not the only anatomical traits that distinguish us from nonhuman primates, even if they may be the most obvious. A list of our most distinctive anatomical features includes the following:

- erect posture and bipedal gait (which entail changes of the backbone, hipbone, and feet);
- opposing thumbs and arm and hand changes (making possible precise manipulation);
- reduction of jaws and remodeling of face;
- cryptic ovulation (and extended female sexual receptivity);
- slow development;
- modification of vocal tract and larynx.

Humans are notably different from other animals not only in anatomy, but also and no less importantly in their behavior, both as individuals and socially. A list of distinctive human behavioral traits includes the following:

- intelligence: abstract thinking, categorizing, and reasoning;
- symbolic (creative) language;
- self-awareness and death-awareness;
- tool-making and technology;
- science, literature, and art;
- ethics and religion;
- social organization and cooperation (division of labor);
- legal codes and political institutions.

Humans live in groups that are socially organized, and so do other primates. But primate societies do not approach the complexity of human social organization. A distinctive human social trait is culture, which may be understood as the set of not strictly biological human activities and creations. Culture includes social and political institutions, ways of doing things, religious and ethical traditions, language, common sense and scientific knowledge, art and literature, technology, and in general all the creations of the human mind. The advent of culture has brought with it cultural evolution, a superorganic mode of evolution superimposed on the organic mode, which has, in the last few millennia, become the dominant mode of human evolution. Cultural evolution has come about because of cultural change and inheritance, a distinctively human mode of achieving adaptation to the environment and transmitting it through the generations.

Ethical behavior versus ethical norms

Ethics and ethical behavior may serve as a model case of how we may seek the evolutionary explanation of a distinctively human trait. The objective is to ascertain whether an account can be advanced of ethical behavior as an outcome of biological evolution and, if such is the case, whether ethical behavior was directly promoted by natural selection, or has rather come about as an epigenetic manifestation of some other trait that was the target of natural selection.

The question whether ethical behavior is biologically determined may refer either to (1) the *capacity* for ethics (i.e. the proclivity to judge human actions as either right or wrong) and which I refer to as "ethical behavior," or (2) the moral *norms* or moral codes accepted by human beings for guiding their actions. A similar distinction can be made with respect to language. The issue whether the capacity for symbolic language is determined by our biological nature is different from the question of whether the particular language we speak (English, Spanish, or Japanese) is biologically necessary.

The first question posed asks whether the biological nature of *Homo sapiens* is such that humans are necessarily inclined to make moral judgments and to accept ethical values, to identify certain actions as either right or wrong. Affirmative answers to this first question do not necessarily determine what the answer to the second question should be. Independently of whether humans are necessarily ethical, it remains to be determined whether particular moral prescriptions are in fact determined by our biological nature, or whether they are chosen by society, or by individuals. Even if we were to conclude that people cannot avoid having moral standards of conduct, it might be that the choice of the particular standards used for judgment would be arbitrary or that it depended on some other, nonbiological criteria. The need for moral values does not necessarily tell us what these moral values should be, just as the capacity for language does not determine which language we shall speak.

The thesis that I propose is that humans are ethical beings by their biological nature. Humans evaluate their behavior as either right or wrong, moral or immoral, as a consequence of their eminent intellectual capacities, which include self-awareness and abstract thinking. These intellectual capacities are products of the evolutionary process, but they are distinctively human. Thus, I maintain that ethical behavior is not causally related to the social behavior of animals, including kin and reciprocal "altruism" (Ayala 1987, 1995).

A second thesis is that the moral norms according to which we evaluate particular actions as morally either good or bad (as well as the grounds that may be used to justify the moral norms) are products of cultural evolution, not of biological evolution. The norms of morality belong, in this respect, to the same category of phenomena as the languages spoken by different peoples, their political and religious institutions, and the arts, sciences, and technology. The moral codes are in some respects isomorphic with the biological predispositions of the human species, dispositions we share to some extent with other animals. But this isomorphism between ethical norms and biological tendencies is not necessary or universal: it does not apply to all ethical norms in a given society, much less in all human societies.

This second thesis contradicts the proposal of many distinguished evolutionists who, since Darwin's time, have argued that the norms of morality are derived from biological evolution. It also contradicts the sociobiologists, who have recently developed a subtle version of that proposal. The sociobiologists' argument is that human ethical norms are sociocultural correlates of behaviors fostered by biological evolution. I argue that such proposals are misguided and do not escape the naturalistic fallacy. It is true that both natural selection and moral norms sometimes target the same behavior; that is, the two are consistent. But this consistency between the behaviors promoted by natural selection and those sanctioned by moral norms exists only with respect to the consequences of the behaviors; the underlying causations are completely disparate.

Moral codes, like any other dimensions of cultural systems, depend on the existence of human biological nature and must be consistent with it in the sense that they could not counteract it without promoting their own demise. Moreover, the acceptance and persistence of moral norms is facilitated whenever they are consistent with biologically conditioned human behaviors. But the moral norms are independent of such behaviors in the sense that some norms may not favor, and may hinder, the survival and reproduction of the individual and its genes, which are the targets of biological evolution. Discrepancies between accepted moral rules and biological survival are, however, necessarily limited in scope or would otherwise lead to the extinction of the groups accepting such discrepant rules.

Biological roots of ethical behavior

The question whether ethical behavior is determined by our biological nature must be answered in the affirmative. By "ethical behavior" I mean here to refer to the judging of human actions as either good or bad, which is not the same as "good behavior" (i.e. doing what is perceived as good instead of what is perceived as evil). Humans exhibit ethical behavior by nature because their biological constitution determines the presence of the three necessary conditions for ethical behavior. These conditions are: (1) the ability to anticipate the consequences of one's own actions; (2) the ability to make value judgments; and (3) the ability to choose between alternative courses of action. I shall briefly examine each of these abilities and show that they are consequences of the eminent intellectual capacity of human beings.

The ability to anticipate the consequences of one's own actions is the most fundamental of the three conditions required for ethical behavior. Only if I can anticipate that pulling the trigger will shoot the bullet, which in turn will strike and kill my enemy, can the action of pulling the trigger be evaluated as nefarious. Pulling a trigger is not in itself a moral act; it becomes so by virtue of its relevant consequences. My action has an ethical dimension only if I do anticipate these consequences.

The ability to anticipate the consequences of one's actions is closely related to the ability to establish the connection between means and ends; that is, of seeing a means precisely as means, as something that serves a particular end or purpose. This ability to establish the connection between means and their ends requires the ability to anticipate the future and to form mental images of realities not present or not yet in existence.

The ability to establish the connection between means and ends happens to be the fundamental intellectual capacity that has made possible the development of human culture and technology. A reasonable evolutionary hypothesis to account for this capacity proposes that its roots may be found in the evolution of bipedal gait, which transformed the anterior limbs of our ancestors from organs of locomotion into organs of manipulation. The hands thereby gradually became organs adept for the construction and use of objects for hunting and other activities that improved survival and reproduction.

The construction of tools, however, depends not only on manual dexterity but on perceiving them precisely as tools, as objects that help to perform certain actions, that is, as means that serve certain ends or purposes: a knife for cutting, an arrow for hunting, an animal skin for protecting the body from the cold. The hypothesis I am propounding is that natural selection promoted the intellectual capacity of our biped ancestors because increased intelligence facilitated the perception of tools as tools, and therefore their construction and use, with the ensuing amelioration of biological survival and reproduction.

The development of the intellectual abilities of our ancestors took place over 2 million years or longer, gradually increasing the ability to connect means with their ends and, hence, the possibility of making ever more complex tools serving remote purposes. The ability to anticipate the future, essential for ethical behavior, is therefore closely associated with the development of the ability to construct tools, an ability that has produced the advanced technologies of modern societies and that is largely responsible for the success of humankind as a biological species.

The second condition for the existence of ethical behavior is the ability to make value judgments, to perceive certain objects or deeds as more desirable than others. Only if I can see the death of my enemy as preferable to his or her survival (or vice versa) can the action leading to his or her demise be thought of in moral terms. If the alternative consequences of an action are neutral with respect to value, the action does not belong within the scope of ethical behavior. The ability to make value judgments depends on the capacity for abstraction, that is, on the capacity to perceive actions or objects as members of general classes. This makes it possible to compare objects or actions with one another and to perceive some as more desirable than others. The capacity for abstraction, necessary to perceive individual objects or actions as members of general classes, requires an advanced intelligence such as exists only in humans. Thus, I see the ability to make value judgments primarily as an implicit consequence of the enhanced intelligence favored by natural selection in human evolution. Nevertheless, valuing certain objects or actions and choosing them over their alternatives can be of biological consequence; doing this in terms of general categories can be beneficial in practice.

Value judgments indicate preference for what is perceived as good and rejection of what is perceived as bad; good and bad may refer to economic, aesthetic, or all sorts of other kinds of values. Moral judgments concern the values of right and wrong in human conduct. Moral judgments are a particular class of value judgments; namely those where preference is not dictated by one's own interest or profit, but by regard for others, which may cause benefits to particular individuals (altruism), or take into consideration the interests of a social group to which one belongs.

Evolutionists have demonstrated that "group selection" is not an "evolutionary stable strategy." Group selection refers to selection that benefits the group at the expense of the (inclusive) fitness of the individual. Suppose that there is a group with a genetic trait that benefits the group to the extent that the group is very successful and expands in numbers at the expense of other groups and to the benefit and multiplication of the individuals in the group and their genetic make-ups. Suppose now that a mutation arises in an individual that makes it behave selfishly. Individuals carrying this mutation will benefit from the altruistic behavior of the others and will not incur the costs of the others' altruistic behavior. Consequently the selfish individuals will have higher fitness than the altruists and the selfish mutation will increase in frequency until it eliminates from the group the altruistic gene.

Humans, however, can perceive the benefits of altruistic behavior for the group (and through the group to themselves) and choose to behave altruistically. The altruistic behavior may be enforced by political authority imposing a penalty (if you commit adultery, or if you steal, you'll be stoned to death, or jailed, or otherwise punished) or promoted through religious authority or belief, like the Christian commandments against adultery and theft. Thus, morality makes it possible for true altruism to be an evolutionary stable strategy. But this depends on humans' exalted intelligence and the presence of the three conditions for moral behavior.

The third condition necessary for ethical behavior is the ability to choose between alternative courses of action. Pulling the trigger can be a moral action only if I have the option not to pull it. A necessary action beyond our control is not a moral action: the circulation of the blood or the digestion of food are not moral actions.

Whether there is free will has been much discussed by philosophers and this is not the appropriate place to review the arguments. I will only advance two considerations based on common-sense experience. One is our profound personal conviction that the possibility of choosing between alternatives is genuine rather than only apparent. The second consideration is that when we confront a given situation that requires action on our part, we are able mentally to explore alternative courses of action, thereby extending the field within which we can exercise our free will. In any case, if there were no free will, there would be no ethical behavior; morality would only be an illusion. The point that I wish to make here is, however, that free will is dependent on the existence of a well-developed intelligence, which makes it possible to explore alternative courses of action and to choose one or another in view of the anticipated consequences.

In summary, ethical behavior is an attribute of the biological make-up of humans and is, in that sense, a product of biological evolution. But I see no evidence that ethical behavior developed because it was adaptive in itself. I find it hard to see how *evaluating* certain actions as either good or evil (not just choosing some actions rather than others, or evaluating them with respect to their practical consequences) would promote the reproductive fitness of the evaluators. Nor do I see how there might be some form of "incipient" ethical behavior that would then be further promoted by natural selection.

It seems rather that the likely target of natural selection was the development of advanced intellectual capacities. This development was favored by natural selection because the construction and use of tools improved the strategic position of our biped ancestors. Once bipedalism evolved and tool-using and tool-making became possible, those individuals more effective in these functions had a greater probability of biological success. The biological advantage provided by the design and use of tools persisted long enough so that intellectual abilities continued to increase, eventually yielding the eminent development of intelligence that is characteristic of *Homo sapiens*.

Ethical norms: beyond biology

Since the publication of Darwin's theory of evolution by natural selection, philosophers as well as biologists have attempted to find in the evolutionary process the justification for moral norms. The common ground in such proposals is that evolution is a natural process that achieves goals that are desirable and thereby morally good; indeed it has produced humans (Ayala 1987). Proponents of these ideas claim that only the evolutionary goals can give moral value to human action: whether a human deed is morally right depends on whether it directly or indirectly promotes the evolutionary process and its natural objectives.

A different attempt to ground moral codes on the evolutionary process is that of the sociobiologists, particularly from E. O. Wilson (1975, 1978; see also Alexander 1987), who starts by proposing that "scientists and humanists should consider together the possibility that the time has come for ethics to be removed temporarily from the hands of the philosophers and biologicized" (Wilson 1975, 562). The sociobiologists argue that the perception that morality exists is an epigenetic manifestation of our genes, which so manipulate humans as to make them believe that some behaviors are morally "good" so that people behave in ways that are good for their genes. Humans might not otherwise pursue these behaviors (altruism, for example) because their genetic benefit is not apparent (except to sociobiologists after the development of their discipline) (Ruse 1986a, 1986b; Ruse and Wilson 1986).

Wilson writes: "Human behavior – like the deepest capacities for emotional response which drive and guide it – is the circuitous technique by which human genetic material has been and will be kept intact. *Morality has no other demonstrable ultimate function*" (Wilson 1978, 167, my italics). How is one to interpret this statement? It is possible that Wilson is simply giving the reason why ethical behavior exists at all, in the sense I have just stated; namely, our genes prompt us to accept what we call "morality," so that we act accordingly to the interests of our genes, interests that are not otherwise apparent to us.

It is possible, however, to read Wilson's statement as a justification of human moral codes: the function of these would be to preserve human genes. But this would entail the naturalistic fallacy and, worse yet, would seem to justify a morality that most of us detest. If the preservation of human genes (be it those of the individual, the group, or the species) is the purpose that moral norms serve, Spencer's Social Darwinism would seem right; racism or even genocide could be justified as morally correct if they were perceived as the means to preserve those genes thought to be good or desirable and to eliminate those thought to be bad or undesirable. There is no doubt in my mind that Wilson is not intending to justify racism or genocide, but this is one possible interpretation of his words.

I shall now turn to the sociobiologists' proposition that natural selection favors behaviors that are isomorphic with the behaviors sanctioned by the moral codes endorsed by most humans.

Evolutionists had for years struggled with finding an explanation for the apparently altruistic behavior of animals. When predators attack a herd of zebras, these will attempt to protect the young in the herd, even if they are not their progeny, rather than fleeing. When a prairie dog sights a coyote, it will warn other members of the colony with an alarm call, even though by drawing attention to itself this increases its own risk. Examples of altruistic behaviors of this kind can be multiplied.

Altruism is defined in the dictionary I happen to have at hand (*Merriam Webster's Collegiate Dictionary*, 10th edn.) as "unselfish regard for, or devotion to the welfare of others." The dictionary gives a second definition: "behavior by an animal that is not beneficial to or may be harmful to itself but that benefits others of its species." To speak of animal altruism is not to claim that explicit feelings of devotion or regard are present in them, but rather that animals act for the welfare of others at their own risk just as humans are expected to do when behaving altruistically. The problem is precisely how to justify such behaviors in terms of natural selection. Assume, for illustration, that in a certain species there are two alternative forms of a gene ("alleles"), of which one but not the other promotes altruistic behavior. Individuals possessing the altruistic allele will risk their life for the benefit of others, whereas those possessing the non-altruistic allele will benefit from altruistic behavior without risking themselves. Possessors of the altruistic allele will be more likely to die and the allele will therefore be eliminated more often than the non-altruistic allele. Eventually, after some generations, the altruistic allele will be completely replaced by the non-altruistic one. But then how is it that altruistic behaviors are common in animals without the benefit of ethical motivation?

One major contribution of sociobiology to evolutionary theory is the notion of "inclusive fitness." In order to ascertain the consequences of natural selection it is necessary to take into account a gene's effects not only on a particular individual but on all individuals possessing that gene. When considering altruistic behavior, one must take into account not only the risks for the altruistic individual, but also the benefits for other possessors of the same allele. Zebras live in herds where individuals are blood relatives. A gene prompting adults to protect the defenseless young would be favored by natural selection if the benefit (in terms of saved carriers of that gene) is greater than the cost (due to the increased risk of the protectors). An individual that lacks the altruistic gene and carries instead a non-altruistic one, will not risk its life, but the non-altruistic allele is partially eradicated with the death of each defenseless relative.

It follows from this line of reasoning that the more closely related the members of a herd or animal group are, the more altruistic behaviors should be present. This seems to be generally the case. We need not enter here into the details of the quantitative theory developed by sociobiologists in order to appreciate the significance of two examples. The most obvious is parental care. Parents feed and protect their young because each child has half the genes of each parent: the genes are protecting themselves, as it were, when they prompt a parent to care for its young.

A second example is more subtle: the social organization and behavior of certain animals like the honeybee. Worker bees toil building the hive and feeding and caring for the larvae even though they themselves are sterile and only the queen produces progeny. Assume that in some ancestral hive, a gene arises that prompts worker bees to behave as they now do. It would seem that such a gene would not be passed on to the following generation because such worker bees do not reproduce. But such inference is erroneous. Queen bees produce two kinds of eggs: some that remain unfertilized develop into males (which are therefore "haploid," i.e. carry only one set of genes); others that are fertilized (hence, are "diploid," i.e. carry two sets of genes) develop into worker bees and occasionally into a queen. W. D. Hamilton (1964) demonstrated that with such a reproductive system daughter queens and their worker sisters share two-thirds of their genes, whereas daughter queens and their mother share only one half of their genes. Hence, the worker bee genes are more effectively propagated by workers caring for their sisters than if they were to produce and care for their own daughters. Natural selection can thus explain the existence in social insects of sterile castes, which exhibit a most extreme form of apparently altruistic behavior by dedicating their life to care for the progeny of another individual (the queen). The theory predicts that the hive will tend to minimize the number of reproductive females, which is what happens in the honeybee, where all reproduction is performed by the one queen.

Sociobiologists point out that many of the moral norms commonly accepted in human societies sanction behaviors also promoted by natural selection (which promotion becomes apparent only when the inclusive fitness of genes is taken into account). Examples of such behaviors are the commandment to honor one's parents, the incest taboo, the greater blame attributed to the wife's than to the husband's adultery, the ban or restriction on divorce, and many others. The sociobiologists' argument is that human ethical norms are sociocultural correlates of behaviors fostered by biological evolution. Ethical norms protect such evolution-determined behaviors as well as being specified by them.

I believe, however, that the sociobiologists' argument is misguided and does not escape the naturalistic fallacy (Ayala 1987, 1995; see also Sober and Wilson 1998). Consider altruism as an example. Altruism in the biological sense (altruism$_b$) is defined in terms of the population genetic consequences of a certain behavior. Altruism$_b$ is explained by the fact that genes prompting such behavior are actually favored by natural selection (when inclusive fitness is taken into account), even though the fitness of the behaving individual is decreased. But altruism in the moral sense (altruism$_m$) is explained in terms of motivations: a person chooses to risk his own life (or incur some kind of "cost") for the benefit of somebody else. The isomorphism between altruism$_b$ and altruism$_m$ is only apparent: an individual's chances are improved by the behavior of another individual who incurs a risk or cost. The underlying causations are completely disparate: the ensuing genetic benefits in altruism$_b$; regard for others in altruism$_m$. (Sociobiologists, however, might say that our perception that our altruistic behavior is motivated by regard for others is itself caused by our genes that seek that way to accomplish their own purposes. As Ruse [1986b] puts it: "[Sociobiologists] argue that moral (literal) altruism might be one way in which biological (metaphorical) 'altruism' could be achieved … Literal, moral altruism is a major way in which advantageous biological cooperation is achieved … In order to achieve [biological] 'altruism' we are altruistic." This is, of course, a claim of biological determinism in the extreme, and ultimately entails the denial of true free will.)

One additional observation worth notice is that some norms of morality are consistent with behaviors prompted by natural selection, but other norms are not so. The commandment of charity, "Love thy neighbor as thyself," often runs contrary to the inclusive fitness of the genes, even though it promotes social cooperation and peace of mind. If the yardstick of morality were the multiplication of genes, the supreme moral imperative would be to beget the largest possible number of children and (with lesser dedication) to encourage our close relatives to do the same. But to impregnate the most women possible is not, in the view of most people, the highest moral duty of a man.

As I stated above, my view is that we make moral judgments as a consequence of our eminent intellectual abilities, not as an innate way for achieving biological gain, and that the codes of morality by which humans guide their actions have been formulated as a consequence of social traditions and/or by religious or political authority (Wilson 2002). The codes of morality that prevail in human populations are largely consistent with the genetic interests of individuals because otherwise the codes would not have historically survived. But the codes also incorporate norms that benefit the tribe or community rather than the individual, such as the commandments against adultery or theft.

I summarize my views by reference to the analogy between ethical behavior and human languages. Our biological nature determines the sounds that we can or cannot utter and also constrains human language in other ways. But a language's syntax and vocabulary are not determined by our biological nature (otherwise there could not be a multitude of tongues), but are products of human culture. Likewise, moral norms are not determined by biological processes, but by cultural traditions and principles, including religious beliefs, that are products of human history.

REFERENCES

Alexander, R. D. (1987) *The Biology of Moral Systems*. Hawthorne, NY: Aldine.

Avise, J. C. and F. J. Ayala (eds.) (2009) *In the Light of Evolution*, vol. III, *Two Centuries of Darwin*. Washington, DC: National Academies Press.

Ayala, F. J. (1968) "Biology as an Autonomous Science." *American Scientist* 56 (3): 207–21.

(1987) "The Biological Roots of Morality." *Biology and Philosophy* 2: 235–52.

(1995) "The Difference of Being Human: Ethical Behavior as an Evolutionary Byproduct." In H. Rolston, III (ed.), *Biology, Ethics, and the Origin of Life*. Boston and London: Jones and Bartlett, pp. 113–35.

(1997) "Ascent by Natural Selection. Review of *Monad to Man*, by M. Ruse." *Science* 275: 495–96.

Ayala, F. J. and R. Arp (eds.) (2010) *Contemporary Debates in Philosophy of Biology*. Malden, MA: Wiley-Blackwell.

Ayala, F. J. and T. Dobzhansky (eds.) (1974) *Studies in the Philosophy of Biology: Reduction and Related Problems*. Berkeley, CA: University of California Press.

Green, R. E., J. Krause, A. W. Briggs, T. Maricic, U. Stenzel, M. Kircher *et al.* (2010) "A Draft Sequence of the Neandertal Genome." *Science* 328 (5979): 710–22.

Hamilton, W. D. (1964) "The Genetic Evolution of Social Behavior." *Journal of Theoretical Biology* 7: 1–52.

Hammer, M. F., A. E. Woerner, F. L. Mendez, J. C. Watkins, and J. D. Wall (2011) "Genetic Evidence for Archaic Admixture in Africa." *Proceedings of the National Academy of Sciences* 108 (37): 15123–28.

Meyer, M., M. Kircher, M.-T. Gansauge, Heng Li, F. Racimo, S. Mallick *et al.* (2012) "A High-Coverage Genome Sequence from an Archaic Denisovan Individual." *Science* 338 (6104): 222–26.

Rasmussen, M., Xiaosen Guo, Yong Wang, K. E. Lohmueller, S. Rasmussen, A. Albrechtsen *et al.* (2011) "An Aboriginal Australian Genome Reveals Separate Human Dispersals into Asia." *Science* 334 (6052): 94–98.

Reich, D., R. E. Green, M. Kircher, J. Krause, N. Patterson, E. Y. Durand *et al.* (2010) "Genetic History of an Archaic Hominin Group from Denisova Cave in Siberia." *Nature* 468: 1053–60.

Ruse, M. (1986a) *Taking Darwin Seriously: A Naturalistic Approach to Philosophy.* Oxford: Basil Blackwell.

(1986b) "Evolutionary Ethics: A Phoenix Arisen." *Zygon* 21: 95–112.

(1988) *Philosophy of Biology Today.* Albany NY: SUNY Press

(2012) *The Philosophy of Human Evolution.* Cambridge University Press.

Ruse, M. and E. O. Wilson (1986) "Moral Philosophy as Applied Science." *Journal of the Royal Institute of Philosophy* 61: 173–92.

Skoglund, P. and M. Jakobsson (2011) "Archaic Human Ancestry in East Asia." *Proceedings of the National Academy of Sciences* 108: 18301–06.

Sober, E. and D. S. Wilson (1998) *Unto Others: The Evolution and Psychology of Unselfish Behavior.* Cambridge, MA: Harvard University Press.

Stringer, C. (2012) "Evolution: What Makes a Modern Human." *Nature* 485: 33–35.

Wilson, D. S. (2002) *Darwin's Cathedral: Evolution, Religion, and the Nature of Society.* University of Chicago Press.

Wilson, E. O. (1975) *Sociobiology: The New Synthesis.* Cambridge, MA: Harvard University Press.

(1978) *On Human Nature.* Cambridge, MA: Harvard University Press.

PART I

Evolution and theology

Evolutionary theory, causal completeness, and theism
The case of "guided" mutation

Elliott Sober

Michael Ruse (2004) and I (Sober 2010, 2011) are both "accommodationists." We think that evolutionary theory and some types of theism can be reconciled. We also have in common the negative fact that neither of us is a theist.

It is obvious that some kinds of theism are logically inconsistent with evolutionary theory; it is equally obvious that other kinds of theism are logically compatible with that theory. Young Earth Creationism says that life on Earth began some 10,000 to 50,000 years ago by God's separately creating each species (or each "basic kind" of organism). Evolutionary theory says that life began about 3.8 billion years ago and that all current species are genealogically related. There is no reconciling evolutionary theory with this form of theism; one must be wrong if the other is right. At the other end of the spectrum is deism, which is the view that God created (1) the universe, (2) the laws of nature, and (3) the initial conditions of the universe, and then sat back, allowing everything that happens in nature to be a consequence of those three items. Evolutionary theory fails to conflict with this form of theism for the simple reason that the theory says nothing at all about the origin of the universe, or its initial state, or where the laws of nature come from.

The kind of theism that interests me lies in between these two extremes. I am interested in interventionist[1] theisms that say that God not only

I gave lectures on this material at Florida State University, the University of Chicago, the University of Connecticut at Storrs, and the University of Cambridge. I am grateful to members of these audiences for useful discussion. I also thank Steve Nadler and Mike Steel for their comments.
[1] I am using the term "intervention" in a way that is broader than what many theologians prefer. They often use the word to name only the piecemeal local effects that God has concerning the details of some small aspect of the universe's total history. They prefer to use the term "interaction" for divine causation that influences large-scale, global features of the universe. If God parted the Red Sea, this would be for them an intervention in the narrow sense; if God sustains the whole universe from each moment to the next, this is an example of an interaction, not an intervention. I am using the term "intervention" in a broader sense, so that it applies both to narrow-sense interventions and to interactions.

produced the (1)–(2)–(3) that Deism describes, but also intervened in nature after the universe's beginning. Traditional Judaism, Christianity, and Islam (if I may use the loose term "traditional") are theisms of this sort. My question is whether a God that intervenes in human history, and in nature more generally, can be reconciled with evolutionary theory. I won't discuss all of the interventions that an interventionist might want to endorse. Rather, I'll limit myself to the idea that God intervenes in the evolutionary process by causing this or that mutation to occur at a given time and place.[2] I say that I am interested in interventionist theisms, but this is not because I believe any of them (recall my first paragraph). I am interested in them because many religious people accept interventionist theisms and think that this obliges them to reject evolutionary theory. Biologists say that they have abundant evidence that mutations are unguided. This seems to mean that God does not intervene in the evolutionary process, at least not by causing this or that mutation to occur. I'll argue that what biologists mean, or ought to mean, when they say that mutations are unguided says nothing about whether God ever causes a mutation to occur.

The argument I'll give for thinking that evolutionary theory is logically compatible with this type of divine intervention is simple; it relies just on the fact that evolutionary theory, properly understood, is a probabilistic theory. The argument therefore generalizes; it applies to pretty much *any* probabilistic theory, not just to evolutionary theory. What I'll say in what follows expands on what I have said in Sober (2010, 2011) by filling in some details and by replying to some objections. Although I'll argue that evolutionary theory, properly understood, does not rule out God's causing some mutations, the theory does rule this out when you add something to it. But the something else is a philosophical thesis, not a scientific theory at all.

Evolutionary theory and determinism

Evolutionary theory, in its application to finite populations of organisms, is a probabilistic theory. The theory does not tell you what must happen in the future, given a description of the population's present state. Rather,

---

[2] This idea has a history, beginning with Darwin's relationship with the Harvard botanist Asa Gray. Gray was Darwin's foremost advocate in North America, but he urged Darwin (both in print and in their extensive and congenial private correspondence) to add "in the philosophy of his hypothesis, that variation has been led along certain beneficial lines" (Gray 1888). Gray had God in mind as the agent that was doing the guiding. Darwin always demurred. It is worth mentioning that both deists and interventionists can maintain that God has arranged for this or that mutation to occur; the difference is that deists think that God does this indirectly while interventionists thinks that God acts more directly.

it tells you that different futures are possible and assigns a probability to each. This fact about the theory isn't immediately obvious when you look at various simple mathematical models of the evolutionary process. For example, consider a haploid population in which there are just two alleles A and B at a locus that have frequencies p and q and constant fitnesses $w(A)$ and $w(B)$. The change in trait frequency in the next generation is (Crow and Kimura 1970, 179):

$$\Delta p = \frac{spq}{\bar{w}}$$

The selection coefficient s represents the fitness difference $[w(A) - w(B)]$ and $\bar{w}$ is the average fitness of individuals in the parental generation. This equation says what will happen, not just what will probably happen. This is because the model describes an infinite population. As soon as you take account of the fact that real populations are finite, you need to reinterpret this equation; it tells you about the *mathematical expectation* of the change in trait frequency. The fitnesses (and so the selection coefficient) are probabilistic quantities; in a finite population there are different possible changes in trait frequency that might ensue, each with its own probability; the expected value of Δp is the probabilistically weighted *average* of the changes that might ensue.

What does the probabilistic character of evolutionary theory (for the case in which populations are finite) tell you about whether determinism is true? Determinism is the following thesis:

> Determinism: A complete description of the history of the universe up to time t uniquely determines what the future of the universe will be after time t.

If you assume that the history of the universe has the Markov property, this formulation can be replaced with something that is logically stronger:

> Determinism$_M$: A complete description of the universe at time t uniquely determines what the future of the universe will be after time t.

In both formulations, a complete description of the past leaves open just one possible future; this is the one that must happen. Indeterminism, on the other hand, says that a complete description of the past leaves open multiple possible futures, each of which has its own probability. I'll use the Markovian formulation of determinism in what follows.

To see what evolutionary theory says about determinism, let's switch to an easier question. When we toss a coin we usually assume that the coin is fair, meaning that

(1) Pr(the coin lands heads at t_2 | the coin was tossed at t_1) = 0.5.

Notice that this conditional probability describes a relationship between two propositions, the ones that flank the conditional probability sign " | ". Given just that you toss the coin at t_1, the probability of getting heads at t_2 is 1/2. Although proposition (1) is true (let us suppose), it says nothing about whether determinism is true because (1) does not say, one way or the other, whether "the coin was tossed at t_1" is a complete description of what is true at that time. It is perfectly compatible with proposition (1) that the following is true:

(2) Pr(the coin lands heads at t_2 | a complete description of the state of the coin-tossing set-up at t_1) = 1.0.

Proposition (1) also is logically compatible with the probability described in (2) having an intermediate value (in which case determinism would be false). Do not make the mistake of thinking that there is one true probability that the coin has of landing heads at t_2 and that therefore (1) and (2) are in conflict. Probability is like distance. There is no such thing as the one true distance to Madison; there's the distance from Omaha to Madison and also the distance from Atlanta to Madison. Probability, like distance, is an inherently relational concept (Sober 2011).[3]

If (1) and (2) are both true, (1) provides a *causally incomplete* representation of the coin-tossing system. By this I don't mean that (1) fails to mention all of the many causes that affect the outcome of this coin toss, a set of events that traces back into the past all the way to the origin of the universe. Rather, what I mean by causal incompleteness is this:

> X at t_1 is a causally incomplete description of whether Y is true at t_2 if and only if there exists a proposition H that is true at t_1 such that Pr(Y at t_2 | X at t_1) ≠ Pr(Y at t_2 | X at t_1 & H at t_1).[4]

If (1) is true but causally incomplete, there is a "hidden variable" (H). This means that there is a causally relevant factor at t_1 that is "hidden" in the

---

[3] Renyi (1970, 34) says that "every probability is in reality a conditional probability" and puts this sentence in italics, for emphasis. Renyi's idea was that "conditional probability" should be taken as primitive, and that a concept of unconditional probability could then be derived from it. Current practice is mainly to do the opposite. My idea that probabilities are relational pertains to both conditional and to so-called unconditional probabilities. Unconditional probabilities make sense when they are based on a model; $Pr_M(A)$ involves a relation between propositions just as much as $Pr_M(A | B)$ does. My thesis of relationality has an exception: tautologies and contradictions can be said to have their probabilities (0 and 1, respectively) nonrelationally.
[4] I assume that both of the probabilities in this inequality are well defined. Note that if determinism is true, then a causally complete model must be deterministic, whereas if determinism is false, then

sense that it goes unmentioned in (1). Probability statements like (1) can be true without being causally complete.[5]

There is nothing special about this coin-tossing example. *Any* probability statement you please (perhaps with the exception of the laws of quantum mechanics, which raise special questions that I won't address here) can be true without being causally complete. Evolutionary theory, I've suggested, is a probabilistic theory. This means that it can be true without being causally complete. The theory doesn't rule out the possibility that there are hidden variables. This means that it, like many other probability statements, doesn't rule out the possibility that there are *supernatural* hidden variables.[6]

Guided mutations

Biologists now know vastly more about the mutation process than Darwin did. They often summarize this knowledge by saying that mutations are "unguided." This seems to entail that no one guides mutations, not even God. Here I'll explain why our scientific knowledge of mutation, properly understood, does not entail that God never guides mutations.

In his book *The Variation of Animals and Plants under Domestication*, Darwin (1868, 249), without any knowledge of Mendelism, managed to characterize what remains the modern understanding of the biological idea that mutations are unguided:

> Let an architect be compelled to build an edifice with uncut stones, fallen from a precipice. The shape of each fragment may be called accidental; yet the shape of each has been determined by the force of gravity, the nature of the rock, and the slope of the precipice, – events and circumstances all of which depend on natural laws; but there is no relation between these

a causally complete model cannot be deterministic and still be true. I also should mention that the concept of causal completeness (like the concept of determinism itself) requires one to be careful about how one understands a description of what is true "at at a given time." If you pack the whole future of the world into what you call "a description of the system at time t_1," then determinism is trivially true. This logical trick should not be taken to show that the question of determinism is silly. This point about choice of descriptors is made by Earman (1986), among others.

[5] It is interesting that a deterministic theory, if true, must be causally complete. Suppose that if $C_1, C_2, \ldots,$ C_n occur at t_1, then E must occur at t_2. It follows that $\Pr(E \text{ at } t_2 \mid C_1, C_2, \ldots, C_n \text{ at } t_1) = 1$. Since probabilities of 0 and 1 are "sticky," there can't be a hidden variable at t_1 that, when included in the conditioning proposition, confers on E at t_2 a probability that differs from 1. In contrast, an indeterministic theory can be true without being causally complete.

[6] My argument has points of contact with ideas that Russell (2008) developed for a view he calls NODA (noninterventionist objective divine action); one difference is that he defends the existence of supernatural hidden variables, while I am merely noting their compatibility with evolutionary theory.

laws and the purpose for which each fragment is used by the builder. In the
same manner the variations of each creature are determined by fixed and
immutable laws; but these bear no relation to the living structure which is
slowly built up through the power of selection, whether this be natural or
artificial selection.

"Unguided" does not mean *uncaused*. What it does mean is that mutations do not arise because they would benefit the organisms in which
they occur. Once again, an analogy with gambling devices is apt. Coins
land heads or tails when tossed and these outcomes have their causes, but
one thing that does not influence the outcomes is that gamblers have bet
on them.

To understand more precisely what is involved in the biological claim
that mutations are unguided, I want to consider a simple experiment (Sober
2011). Imagine a species of blue organisms; we'll take a large number of them
and put them in a red environment and an equally large number of those
blue organisms and put them in a green environment. We will monitor how
often those blue organisms mutate to red and how often they mutate to
green in each of the two environments, thus obtaining the data shown in
Table 2.1. The experiment involves a large number of blue organisms so that
we get mutation frequencies different from zero (mutation probabilities are
small). And there is one more detail: these organisms gain a selective advantage from matching their environments. Red organisms outsurvive green
organisms in a red environment, but the reverse is true in a green environment. This might be because predators use color vision to hunt for the
organisms we're talking about, or because human experimenters follow the
protocol of allowing organisms that match their environment to survive and
reproduce while preventing non-matching organisms from doing so.

What does a model of guided mutations say about this experiment? A
model of this type will describe how the probabilities of various mutations
are influenced by what would be good for the organisms in question. The
model proposes the following two probabilistic inequalities:

(G)   Pr(red mutation | red environment) >
                          Pr(red mutation | green environment)
        Pr(red mutation | red environment) >
                          Pr(green mutation | red environment).

These two inequalities are logically independent of each other. The first
says that blue organisms have a higher probability of mutating to red in
a red environment than they have of doing so in a green environment.
The second says that, in a red environment, red mutations have a higher

Table 2.1 *The frequencies with which blue organisms mutate to red and to green in each of two environments*

	Red environment	Green environment
Red mutation	f_1	f_2
Green mutation	f_3	f_4

Table 2.2 *The model of guided mutations entails four probabilistic inequalities. Below are probabilities of the form* Pr(mutation | environment)

	Red environment		Green environment
Red mutation	p_1	>	p_2
	V		∧
Green mutation	p_3	<	p_4

probability of arising than do green mutations. Both inequalities are needed to express what the idea of guided mutation means. The first, by itself, could be true just because red environments are more mutagenic than green environments. The second, by itself, could be true just because red mutations are more probable than green ones, regardless of the environment. To complete our statement of what the guided model says about our experiment, we need to add two further inequalities:

Pr(green mutation | green environment) >
  Pr(green mutation | red environment)
Pr(green mutation | green environment) >
  Pr(red mutation | green environment).

Table 2.2 summarizes the four inequalities that the model of guided mutations proposes. Imagine that each of the four squares in the 2-by-2 table has a height above the surface of the page that represents how big the probability is of the mutation it describes. The model of guided mutation says that there are two peaks and two valleys in Table 2.2.

What does a model that *denies* guided mutation say about our experiment? There are many such models; each of them denies one or more of the four inequalities displayed in Table 2.2. The simplest of all these deniers is a null model. It says that all the probabilities in Table 2.2 are equal:

(N)   Pr(red mutation | red environment) =
                        Pr(red mutation | green environment)
          Pr(red mutation | red environment) =
                        Pr(green mutation | red environment)
          Pr(green mutation | green environment) =
                        Pr(green mutation | red environment)
          Pr(green mutation | green environment) =
                        Pr(red mutation | green environment)

I use the term "null" because this model says there is no difference between any two of these probabilities.

What is the result of our experiment? What won't happen, if this experiment is like the many more sophisticated experiments that biologists have carried out, is that the frequencies described in Table 2.1 exhibit inequalities that "mirror" the inequalities among probabilities depicted in Table 2.2 (where the differences among these observed frequencies are statistically significant). Biologists will conclude from these non-mirroring observations that the guided model (G) is inferior to some model or other that disagrees with (G); perhaps the null model (N) turns out to be the best of these alternatives.[7]

It is experiments and observations of the sort just described that underlie the conviction that biologists have that mutations are "unguided." Suppose the null model (N) is true for the experiment we've been considering. The point to notice here is that the equalities stated in (N) can be true without any of the probabilities mentioned in (N) being *causally complete*. Consider, for example, one of the probabilities discussed by the null model, Pr(red mutation | red environment). There is no reason to think that the occurrence of a red mutation in our blue organisms isn't influenced by causal factors that go unmentioned in this probability. For example, maybe ambient temperature is relevant, with the consequence that

          Pr(red mutation | red environment) ≠
          Pr(red mutation | red environment & hot).

---

[7] Although fit-to-data is an important consideration in evaluating these models, there is something more – the number of adjustable parameters the different models contain. Model selection criteria like the Akaike Information Criterion take both properties of the model into account (see Sober 2008 for further discussion). Notice that the null mode (N) has fewer adjustable parameters than the guided model (G).

The null model (N) does not rule out that there may be hidden variables. So it doesn't rule out that there may be *supernatural* hidden variables.

What, then, are we to make of the considerable evidence that biologists have amassed against guided mutations? They have checked numerous organisms in numerous environments and have monitored the frequencies of numerous different mutations. In every case, a model like (G) turns out to be inferior to a model that denies that mutations are guided. It is unobjectionable to generalize from these consistent findings. So let us suppose that the experiment and the results I described are entirely typical. This means that when you take up a new organism and consider other possible mutations, you should expect that a model like (G) will not be your best model of how mutation probabilities are related to each other. Your model for this new organism and this new set of mutations will be like the old one I described for our blue organisms. The new model won't say of itself that it is causally complete any more than our model for the blue organisms did. These models, old and new, describe the effect of manipulating an organism's natural environment and how those manipulations affect (or fail to affect) mutation probabilities. None of these models rules out hidden variables. So none rules out *supernatural* hidden variables. Just as a model can be true without being causally complete, so too can a model be both true and inductively generalizable without being causally complete.

Here again, I must repeat the warning I issued at the outset: I am not saying that God intervenes in the mutation process. I am saying that scientific findings should not be misinterpreted.

Two ways to think about conditional probabilities

Suppose you toss a coin repeatedly and use the frequency of heads to estimate the value of the conditional probability Pr(the coin lands heads | the coin is tossed). Perhaps you'll use the procedure of maximum likelihood estimation, assigning to the conditional probability the value that makes the frequency data in your sample most probable; this will mean, if you got 51 heads in 100 tosses, that you'll assign to the conditional probability a value of 0.51.

Now consider a second experiment. You have 100 coins and toss each of them once. You get 51 heads. You want to estimate the probability that a coin sampled at random from these 100 will land heads. You use maximum likelihood estimation and conclude that Pr(the coin lands heads | the coin is tossed) = 0.51.

How should these conditional probabilities be interpreted? One natural interpretation of the first experiment is that each toss of that single coin has a probability of landing heads of 0.51. This is the familiar idea that the coin-tossing process is *i.i.d.* (independent and identically distributed). A natural interpretation of the second experiment is that the coins may differ in their probabilities of landing heads, so what you are in fact estimating in the second experiment is the *average probability* of a coin's landing heads in that population of 100 coins. This interpretation admits the possibility that the coins may in fact have different probabilities of landing heads, but it is compatible with the *i.i.d.* assumption.

How should you interpret the probabilities that figure in the models (G) and (N) concerning our blue organisms? I think they should all be viewed as representing *average* probabilities. Suppose two blue organisms mutate to green among the 10,000 blue organisms that were placed in a green environment in our experiment. Your maximum likelihood estimate is that Pr(green mutation | green environment) = 0.0002. This doesn't mean that the two mutations had exactly the same probability of occurring. The two organisms may have differed in various relevant respects. You should interpret 0.0002 as an *average* probability. In doing so, you are admitting that there may be hidden variables.

What if mutations sometimes were guided (in the biological sense)?

The considerable biological data amassed so far indicate that models like (G) are inferior to models like (N). But suppose that an extraordinary species of organism were discovered that upsets the apple cart. Carefully constructed experiments reveal that these organisms adjust their mutation probabilities along beneficial lines, as (G) describes. That would be an amazing discovery. Would it show that God guides mutations?

Of course not. Although "guided mutation" has been historically connected with the idea that God guides the evolutionary process, this is not because the two ideas are logically connected. The association is a historical accident. The existence of guided mutations in the biologist's sense of that term has no more of a connection with divine intervention than does the idea that some organisms can synthesize vitamin D from sunlight and that others can regenerate lost limbs. In these last two cases, developmental biologists seek to characterize the physical mechanisms in individual organisms that subserve these functions, and evolutionary biologists look for naturalistic explanations for why these traits evolved in ancestral populations. The same naturalistic approach would be set in motion by the hypothetical discovery of "guided" mutations.

If the existence of guided mutations doesn't show that God exists, then the nonexistence of guided mutations doesn't show that God does not exist. Atheists and theists should agree that the biological question is separate from the theological question.

A Duhemian analogy

In his book *The Aim and Structure of Physical Theory*, Pierre Duhem (1954) defended the following two claims about theories in physics:

- Physical theories do not, by themselves, make predictions about what we will observe.
- Physical theories, when supplemented by auxiliary assumptions, do make predictions about what we will observe.

The auxiliary assumptions that Duhem discussed include propositions about the initial and boundary conditions of the system under study and propositions about one's measurement devices. This pair of ideas now goes by the name "Duhem's thesis"; it has had a deep influence on philosophy of science.[8] I suggest that it provides a good model for how biological findings about mutation are related to theistic claims about divine intervention in the mutation process.

Evolutionary theory does not entail that God never intervened in the mutation process, but the theory, when supplemented by auxiliary assumptions, does have implications about divine intervention. Here are some possible auxiliary assumptions. The list is not exhaustive:

> (Deism) God created the universe, the laws that govern the universe, and the initial conditions of the universe, but he never intervenes in natural processes after that first moment.
>
> (The Theology of the Unhidden God) If God ever intervened in the mutation process, then we'd have scientific evidence that mutation probabilities change in beneficial directions when the environment changes.[9]

---

[8] This logical thesis deserves to have the influence it has had, though it does need to be fine-tuned and generalized; see Sober (2008, 144) for discussion. An epistemological thesis is sometimes associated with the logical thesis – that evidence never confirms or disconfirms theories taken by themselves, but only has this impact on conjunctions in which the theories figure. This holistic thesis is mistaken and does not follow from the logical thesis (Sober 2004).

[9] I take my name for this position from Lucien Goldmann (1955), turning his book title on its head. According to Goldmann, the Jansenism of Pascal and Racine asserts that God cannot be known via

(Evidentialism) If you lack scientific evidence as to whether X is true, then you should suspend judgment about whether X is true.

(Fideism) You should believe that God guides the mutation process whether or not you have scientific evidence that he does so.

The first two of these, when added to our best scientific picture of what causes mutation, entail that God never intervened in the mutation process. The third, when conjoined with what biology tells us about mutations (properly understood), entails that we should be agnostic about divine mutational intervention. And the fourth, of course, entails that we should believe that God intervenes in the mutation process. It is beyond the scope of this chapter to consider which of these auxiliary assumptions we should adopt (or whether there are other candidates that are better). My present point is that none of these auxiliaries is part of evolutionary theory; they are – all of them – philosophical theses. My Duhemian claim is that evolutionary theory has consequences about divine intervention in the mutation process only when evolutionary theory is supplemented by further assumptions.

I mentioned at the start that some versions of interventionist theism, like Young Earth Creationism, are logically inconsistent with evolutionary theory. The idea that God sometimes intervenes in the mutation process is different.

Concluding comments

I began this chapter by saying that Michael Ruse and I have our accommodationism in common and that neither of us is a theist. There is a third point of contact. Michael's introductory textbook in philosophy of biology (Ruse 1973) took a sympathetic view of logical positivism. I also have a lot of time for that philosophy (Sober 1999, 2000, 2008, 2010, 2011). Not that I subscribe to the testability theory of meaning. But I do think

the categories that we finite creatures possess, except that we know that God, without cause, confers the gift of grace on some human beings but not on others. Is the theology of the unhidden God the one that the biologist and atheist Jerry Coyne is endorsing when he says: "I think that *the absence of evidence for God, when there should be such evidence*, is indeed empirical evidence against God"? (Italics and underlining his: http://whyevolutionistrue.wordpress.com/2012/05/07/can-god-create-mutations-eliottt-sober-says-we-cant-rule-that-out/.) Perhaps Coyne should write a blog about why Pascal was mistaken to think of God as he did. On the other hand, perhaps Coyne's remark is just a tautology. In any event, the slogan "absence of evidence is evidence of absence" has obvious counterexamples. I do not have evidence that there is a storm on the surface of Jupiter now, but that isn't evidence that there is no such storm (Walton 1996). See Sober (2009a) for discussion of this epistemological principle in a probability framework.

that it is important to distinguish propositions that are testable from those that are not. Young Earth Creationism is testable and false. The proposition that God undetectably intervenes in the evolutionary process is not testable. Positivism says that this proposition is meaningless, but I do not. Neither do I say that it is false; I have no way of telling what its truth value is, and I don't find Ockham's razor an appealing argument for atheism in this instance (Sober 2009b). Scientific theories don't tell you what to think about untestable propositions. That's a philosophical question that needs to be addressed as such.

It is important to distinguish the evidential grounds one has for accepting a proposition from the practical reasons one has for asserting it in public. This chapter has considered accommodationism under the first heading, but I want to close by saying something about the second. I bother to publish in defense of accommodationism in part because I want to take the heat off of evolutionary theory. The more evolutionary theory gets called an atheistic theory, the greater the risk that it will lose its place in public school biology courses in the United States; if the theory is thought of in this way, one should not be surprised if a judge decides that teaching evolutionary theory violates the constitutional principle of neutrality with respect to religion.[10] Indeed, the risk is more profound, since what happens in public education often has ramifications for what happens in the wider culture. Creationists have long held that evolutionary theory is atheistic; defenders of the theory are not doing the theory a favor when they agree. Atheists who think that evolutionary theory provides the beginning of an argument for disbelieving in God should make it clear that their arguments depend on additional premises that are not vouchsafed by scientific theory or data. Philosophy is not a dirty word.

REFERENCES

Crow, J. and M. Kimura (1970) *An Introduction to Population Genetics Theory*. Minneapolis: Burgess Publishing Co.

Darwin, C. (1868) *Variation of Animals and Plants under Domestication*. New York: Appleton (2nd edn., 1876).

Duhem, P. (1954) *The Aim and Structure of Physical Theory*. Princeton University Press (first published 1914, in French).

[10] Here I echo Ruse (2009) when he says "[I]f science generally and Darwinism specifically imply that God does not exist, then teaching science generally and Darwinism specifically runs smack up against the First Amendment."

Earman, J. (1986) *A Primer on Determinism*. Dordrecht: Reidel.

Goldmann, L. (1955) *Le Dieu caché: étude sur la vision tragique dans les "Pensées" de Pascal et dans le théâtre de Racine*. Paris: Gallimard. Translated by P. Thody as *The Hidden God: A Study of Tragic Vision in the "Pensées" of Pascal and the Tragedies of Racine*. London: Routledge, 1964.

Gray, A. (1888) "Natural Selection not Inconsistent with Natural Theology." In *Darwiniana*. Cambridge, MA: Harvard University Press, 1963, pp. 72–145.

Renyi, A. (1970) *Foundations of Probability*. San Francisco: Holden-Day.

Ruse, M. (1973) *The Philosophy of Biology*. London: Hutchinson.

(2004) *Can a Darwinian be a Christian? The Relationship between Science and Religion*. Cambridge University Press.

(2009) "Why I Think the New Atheists are a Bloody Disaster." Biologos Forum: http://biologos.org/blog/why-i-think-the-new-atheists-are-a-bloody-disaster (accessed September 30, 2013).

Russell, R. (2008) *Cosmology from Alpha to Omega: The Creative Mutual Interaction of Theology and Science*. Minneapolis, MN: Fortress Press.

Sober, E. (1999) "Testability." *Proceedings and Addresses of the American Philosophical Association* 73: 47–76.

(2000) "Quine's Two Dogmas." *Proceedings of the Aristotelian Society* 74: 237–80.

(2004) "Likelihood, Model Selection, and the Duhem–Quine Problem." *Journal of Philosophy* 101: 1–22.

(2008) *Evidence and Evolution: The Logic behind the Science*. Cambridge University Press.

(2009a) "Absence of Evidence and Evidence of Absence: Evidential Transitivity in Connection with Fossils, Fishing, Fine-Tuning, and Firing Squads." *Philosophical Studies* 143: 63–90.

(2009b) "Parsimony Arguments in Science and Philosophy: A Test Case for Naturalism$_p$." *Proceedings and Addresses of the American Philosophical Association* 83: 117–55.

(2010) "Evolution without Naturalism." In J. Kvanvig (ed.), *Oxford Studies in Philosophy of Religion 3*. Oxford University Press, pp. 187–221.

(2011) *Did Darwin Write the "Origin" Backwards? Philosophical Essays on Darwin's Theory*. Amherst, NY: Prometheus Books.

Walton, D. (1996) *Arguments from Ignorance*. University Park, PA: Penn State Press.

CHAPTER 3

Religion, truth, and progress

Philip Kitcher

I

During his long and distinguished career, Michael Ruse has shaped the discussion of many issues central to the history and philosophy of biology – indeed, he should share with David Hull the principal credit for founding the field in its contemporary form. All those of us who have followed in the path David and Michael cleared owe both of them a very large debt. That is not to say that Johnny-come-latelys (like me) have always agreed with the proposals of the pioneers. Michael and I have sometimes clashed, in particular when his enthusiasm for biological approaches to human behavior met my concerns about the oversimplifications and confusions of prominent attempts to develop those approaches. At other times, we have stood shoulder to shoulder. Although I too have given time to turning back the challenge of Creationism in its various forms, Michael has been more resolute (and more effective) in his resistance to Creos and Neo-Creos alike. And, more recently, both of us have argued against oversimple reductions of debates about the status of religion. In this chapter, I shall try to articulate a framework for the attitude toward religion we seem to share.

Contemporary atheism usually conceives religion as centered on a core of statements, each of them designed to correspond to the cosmic facts. Judaism, for example, would be viewed as committed to supposing that there was once a man (an ancient pastoralist whom we know as Abraham) who made a covenant (that is, a form of contract in the fully modern sense) with an impressive being of extraordinary power (whose name the faithful are not supposed to write). The most prominent contemporary atheists, the self-styled "Four Horsemen" (Daniel Dennett, Richard Dawkins, Christopher Hitchens, and Sam Harris), attempt to argue that all such doctrinal statements are false. Because they suppose that the acceptance of false ideas about the universe is a Bad Thing, perhaps even the worst predicament that can befall a person, they conclude that religion should be swept away. The case they make is entirely negative: it is no part of the atheist project to explore what might take the place of people's religious attitudes. Like Ruse, I believe this scorched-earth policy to be insensitive to important questions. Although we agree with the Horsemen that there is a sense in which all religious doctrines are false, we don't take this to be the end of the proper discussion of religion.

This reaction strikes many atheists (not only the Horsemen but many of those who ride behind them) as muddled and weak-minded – they decry the qualified profession "I'm an atheist, but …" My aim is to explain, and defend it.

II

Classic anthropological and sociological studies of religion often start in a different place, not with *doctrines* but with *practices*, attending to the multi-dimensional character of religious life.[1] Many scholars who work in this tradition emphasize the enduring importance of religion in human life. They distrust a common grand narrative about the history of Western thought, one in which the abandonment of religion – the turn to "secularism" – is seen as a major progressive step.[2] From their perspective, the criticisms offered by contemporary atheists fail to understand the most important features of the world's religions, and presuppose a simplistic story about the defeat of religion by science.

To understand and to appraise this view, it is useful to examine a recent major presentation of it. Robert Bellah advocates a "religious pluralism" that takes the "stories" of all the major traditions (the "axial" traditions) seriously, considering them in their own terms as valuable responses to different human situations. What he has in mind is presented most clearly and precisely in his discussion of the physicist Eric Chaisson, who has proposed that the development of the sciences has prepared the way for a new vision of ourselves and the cosmos, "a powerful and true myth."[3] Bellah comments:

[1] Prominent examples are Durkheim 2008 and Evans-Pritchard 1956. A recent version is Robert Bellah's magisterial discussion in his 2012.
[2] Bellah's study is plainly complementary to that undertaken in another long and important book, Charles Taylor's *A Secular Age* (2007). Like Taylor, and like Craig Calhoun, Bellah is suspicious not only of the "grand narrative of secularism," but also of the term itself (for valuable discussions of this, see Taylor 2011 and Calhoun 2011). Their historical points about the shifting usages of 'secular' have inspired my insertion of scare-quotes.
[3] Chaisson 2001, 213. Quoted in Bellah 2012, 47. Future references to *Religion in Human Evolution* will be given parenthetically in the text.

[T]he myth he tells, though it draws on science, is not science and so cannot claim scientific truth. I would argue that the myths told by the ancient Israelite prophets, by Socrates, Plato, and Aristotle, by Confucius and Mencius, and by the Buddha, just to stay within the purview of this book, are all true myths. They overlap with one another and with Chaisson's myth, but even in their conflicts, which are sometimes serious, they are all worthy of belief, and I find it possible to believe in all of them in rather deep but not exclusive ways. (47)

An unsympathetic reader (a Horseman, perhaps) might suppose that a very distinguished sociologist has fallen into extreme confusion, accepting the White Queen's advice to believe six impossible things before breakfast.[4] But Bellah explicitly contrasts the "conflicts" with the idea that there are "ways" of believing the myths in which they are "not exclusive." Moreover, his remarks are prefaced with a deliberate distinction between two different kinds of truth, one applicable to myth, and one to science. I offer the following, possibly over-pedantic, explication of what he appears to have in mind:

(1) There are statements S_1 and S_2 belonging to different myths, such that, interpreted in a particular way (specifically: interpreted as candidates for scientific truth) they would be inconsistent; interpreted in a different way (specifically: interpreted as candidates for mythical truth), it is possible, without contradiction, to affirm both S_1 and S_2 (in fact, S_1 and S_2 are both mythical truths).

To make sense of this, we'll need accounts of scientific truth and mythical truth.

Bellah's own way of bringing his readers to these concepts starts with the idea of multiple realities that he draws from Alfred Schutz. Schutz supposes that we live in a number of different worlds, the "paramount reality" (2) being the "world of daily life." In dreams, in playing or watching sports, in reading or going to movies, we are supposed to enter other realities. The other worlds we visit are typically taken to be "less real than the world of daily life" (3), but they are important to us: "[O]ne of the first things to be noticed about the world of daily life is that *nobody can stand to be in it all the time*" (3). Different cultures construct the world of daily life differently (4), and the worlds they construct can be challenged by art, by science, and most importantly (at least during human history) by religion (4).

Bellah goes on to elaborate these ideas by connecting them with approaches to psychology and to the general understanding of religion,

---

[4] Carroll 2010.

but I think the introduction of "worlds" and "multiple realities" has already introduced conflations that make difficulties for the notions of truth at which he is aiming. To make progress in articulating his insights, we'll need to go more slowly and carefully.

III

"World" is ambiguous. Sometimes we talk of "the world" as that to which we respond in our experience, that we affect by our actions, and, most fundamentally, that in which we are placed. Conceived in this way, "the world" is correlative with "the self" – however the self is identified, the world is the container in which it sits, everything there is apart from it. So far there's only a minimal structure: a privileged object and everything else. Typically, however, we deploy a richer conception of the world, one in which it is constituted in space and time, contains a wide variety of objects, is divided into kinds of things, full of causal relations, and so forth. Schutz's "world of daily life" is a world in this richer sense, and it's entirely correct to suppose that worlds of this sort are partially culturally constituted. As Bellah notes, what counts as a person, a family, or a nation (4) is variable – and so is what counts as motion or as bird. Worlds change when people revise their ideas about the boundaries of objects or about how to classify things together or what is the normal course of a process – Galileo changed the world when he taught us to see a swinging stone as a pendulum.[5] Yet, in a sparer sense, the world remains the same: the containing stuff (relatively unstructured) is as it always was.

Appreciating the two senses of "world" is a major theme in pragmatism, from James and Dewey to Goodman and Rorty and beyond.[6] The fruitful pragmatist thought is that the worlds in which we live – the worlds-of-objects-divided-into-kinds-with-standards-of-normal-functioning-etc – are structured in part by us, in ways that depend both on our psychological faculties and on our purposes. Sapient creatures with different senses would live in different worlds, and people, whose psychological apparatus is constant but whose ends evolve, actually do live in different worlds. *Among* our purposes are those of common-sense inquiry and of the various sciences: we want to find out about bits and pieces[7] of the

[5] The example is from Kuhn's seminal *Structure of Scientific Revolutions* (1962). I explore the extent to which the Kuhnian conception of world-changes can be rehabilitated in Kitcher 2012: ch. 5.
[6] See, for example, James's image of the block of marble that allows itself to be carved in various ways (1981, 274) and Goodman 1978, ch. 1.
[7] As I have repeatedly argued, a complete theory of everything is an impossible illusion. See Kitcher 2001 and 2012.

world (in the relatively unstructured sense) – that is, we want to organize it, in thought and sometimes in physical rearrangement, so that particular parts are ordered, subject to intervention and prediction and understanding. Yet those are not our only purposes. We devise ways of thinking and forms of language directed toward different ends – in play, in literature and art, in ethics, and in religion.

Bellah wants to assimilate these activities and the linguistic practices that accompany them to the structuring of the world that gives rise to the world of daily life and to the revisions induced in developing the various sciences. He is right to suppose that play, art, ethics, and religion all have the *potential* for further world-making, but the Schutz-inspired proliferation of multiple realities is incautious. For there are two kinds of constraints on the structuring of the world, one stemming from us (our faculties, our purposes) and the other from what the world – in the bare, relatively unstructured sense – will let us do in restructuring it. Casual talk of moving from one world to another is especially apt to ignore the second constraint.

It's possible to discuss the activities and their associated linguistic practices in their own terms without making any decision about whether engaging in them constitutes moving to a different world (another "reality"). Recall that the apparent point of the framework is to provide an account of "religious pluralism": we started by trying to fathom the distinction between scientific truth and mythical truth. I propose to explain truth for everyday factual statements and for the statements advanced by the sciences by offering a minimal correspondence theory. For simplicity consider the most elementary paradigm, statements of form "*a* is *F*" (for example: "Socrates is wise").

> (2) "*a* is *F*" is weakly true just in case the entity to which '*a*' refers belongs to the set of entities to which '*F*' refers. Reference is here conceived as a relation between signs and parts of the world (in the bare sense); that relation is not necessarily physical or reducible to physical relations.[8]

The truth conditions for more complex statements are generated in the standard fashion: many-place predicates are treated as referring to sets of ordered sequences of objects, connectives and quantifiers are dealt with by the usual recursion clauses. The qualification of statements as *weakly* true is motivated by the thought that *strong* truth demands a pragmatic constraint:

> (3) A statement is strongly true just in case it is weakly true and the language in which it belongs is apt for the purposes of those who use it.

---

[8] For more detail, see Kitcher 2012, ch. 4.

As we'll see, the separation of the two conditions allows for helpful parallels in considering other types of truth.

Before turning to those other species of truth, I want to show why the quick invocation of "multiple realities" is not helpful. First, consider an instance of one of the practices that might be viewed as taking us beyond the "world of daily life." You pick up one of the volumes in which Conan Doyle wrote about a character much beloved by his readers (although sometimes resented by himself), or you go to a movie about Sherlock Holmes. You consider the statement: "Sherlock Holmes lives in Baker Street." Of course, if you apply the correspondence theory just proposed, and give the constituent terms their standard meaning, that statement will be false. "Sherlock Holmes" doesn't pick out any part of the world (bare sense) and consequently doesn't figure in the set of all those who have ever been residents of Baker Street; "Baker Street" refers to a part of the world, although "221B Baker Street" does not. Yet you might want to count the statement as true, at least "in some sense." You could develop that idea by extending the minimal correspondence theory of truth and invoking another "reality": there's a world, the world of the stories, in which "Sherlock Holmes" does refer and the object to which the term refers lives in the place to which "Baker Street" refers. But what is this "other reality"? How do we "engage with it"? How does it relate to the world (bare sense)?

Here's a much better way to elaborate an account of truth that will vindicate "Sherlock Holmes lives in Baker Street." Our world does[9] contain an author (Conan Doyle) and a number of books written by him. Those books *endorse* certain statements of form "Sherlock Holmes ..." Endorsement is sometimes explicit – one of the books contains the statement itself; sometimes endorsement is a matter of implication.[10] I suggest abandoning a correspondence account in favor of a theory of fictional truth.

> (4) S is weakly fictionally true just in case there's a fiction F consisting of a body of statements, and this body of statements endorses S.

> (5) S is strongly fictionally true just in case S is weakly fictionally true, and the pertinent fiction F is apt for the purposes of those who engage with it.

---

[9] I use the present tense to recognize that, from an atemporal point of view, Conan Doyle is part of our world; the fact that he is now dead is of no significance – and not because of the truth of his spiritualist views!

[10] Deciding which implications count depends on matters of interpretation. By analogy with a famous question posed by L. C. Knights, we could ask how many pipes Sherlock Holmes owned. The books endorse claiming that he owned at least one, although there's no definite number they endorse.

It's worth noting that, for common sense and science, condition (2) is often quite demanding, by contrast with its counterpart, (4). With respect to fictional truth, by contrast, (5), the analog of (3), does the majority of the work of separating truth from falsehood. With respect to fictions (and fictional truth), many are called but few are chosen.

Readers of Conan Doyle are sometimes so taken with his famous detective that they pursue, individually or collectively, all kinds of investigations into details of life in late Victorian London. Few of them are confused into believing that Sherlock Holmes himself will be among the objects of their discovery: they insist on this separation between "the world of daily life" and the "world of the stories." Talk of the latter "world" is dangerous precisely because there is overlap – indeed without the overlap the activities of the Sherlockians would make no sense. When the boundaries of the overlap are left unclear, confusion is easy. An advantage of my approach to fictional truth is that it raises, from the very beginning, the question of how the body of statements constituting the fiction relates to the world of daily life.

Nobody is likely to propose that Conan Doyle's stories are so attuned to our cultural needs, so apt, that they should prompt revision of the world of daily life, to include Sherlock Holmes among its denizens. The general idea that human practices, including art and science and religion, can prompt restructurings of the world is correct, but it is useful to appreciate the limits of our world-making (and remaking). Juxtaposing two scientific cases will bring out the point. The pre-Copernican world was one in which the Earth did not count among the planets; it gave way in the seventeenth century to a world based on different principles of organization, in which the Earth was grouped with Mars, Venus, and the rest, even seen as akin to bodies revolving around other suns. Here it is reasonable to talk of a change in the world of daily life (or, perhaps, in the daily life of the cognoscenti). In the chemical revolution of the late eighteenth century, processes of combustion were reclassified as episodes of absorption rather than of emission. The chemical community abandoned the thesis that things that burn share a common substance (or "principle"), phlogiston, that is given off in combustion. As in the Copernican case, the classification of items in the world changed. But it would be wrong to suppose that the older chemical world contained a substance that had disappeared from its successor. Phlogiston *never* existed. World changes can draw the boundaries of objects in different places, group the same things in different ways, or see processes against the background of a different standard (as Galileo did with the pendulum), but their powers to populate the world with new beings are limited. Defined as "the substance emitted in combustion," the term "phlogiston" always failed to pick out any part of reality, of the world (in the bare sense).

IV

Cavalier talk of multiple realities risks overstepping important limits, confusing world-changes that work from those that don't. Fortunately Bellah's core insights don't depend on the dubious metaphysics he takes over from Schutz. They can be captured by a systematic exploration of types of truth in ways prefigured by the discussions of the last section.

Start with a familiar (but under-appreciated) point from the early sections of Wittgenstein's *Philosophical Investigations*. Our uses of language are quite various, not all to be assimilated to the paradigm of describing or "stating the facts." Yet the diverse language games we play have rules and criteria for playing them successfully, and it is this that underwrites a broad concept of truth spanning many domains. With respect to any given language-game, truth applies to those statements the players of that game aim to achieve.[11] Wittgenstein's repeated warnings against viewing all language as descriptive should be understood as protests against the hegemony of a correspondence theory of truth.

Both in the *Investigations* and in the *Remarks on the Foundations of Mathematics*, Wittgenstein uses the example of mathematics to emphasize the metaphysical peculiarities that result from failure to appreciate the point. If mathematical statements are to count as true, then the supposition that truth is correspondence to reality will quickly generate acceptance of a vast realm of mathematical entities that must serve as the referents of the singular terms mathematicians use ('2', 'the null set', and so on). Many philosophers (and some mathematicians) have acquiesced more or less regretfully in this Platonism, even though it brings in its train awkward questions about how we are able to refer to mathematical objects or to know anything about them.[12] The mysteries, and the troubles they bring, can be avoided entirely if we heed Wittgenstein's warning and think of mathematical truth in a different fashion.

Most mathematicians spend all their lives manipulating symbols within systems they have inherited from others: they play symbolic games whose rules are already laid down. Statements (or formulae) they produce count

---

[11] This point is made by Michael Dummett in his seminal essay "Truth"; see Dummett 1978.
[12] See Benacerraf 1973, and Lear 1977.

as true just in case they can be reached by the established rules. Yet there's a pragmatic condition on the games they play. The games must be worth playing. Some of the systems are very ancient, introduced millennia ago to cope with operations of regrouping objects and matching them, or of laying down units along a stretch of land. Out of the very basic games, those played in arithmetic and geometry, come other questions that call for extended notation and novel systems. The history of mathematics is punctuated by moments in which new symbolic games are introduced, perhaps to cope with physical phenomena, perhaps to answer questions raised by earlier games but unanswerable within them, perhaps to generalize earlier games, perhaps because the new games are aesthetically appealing or simply fun to play.[13] So we have the beginnings of an account of mathematical truth.

> (6) *S* is mathematically true just in case it can be reached by the rules of a symbolic game worth playing.
>
> (7) Symbolic games are worth playing just in case they contribute to inquiries into physical reality or answer questions raised by previous symbolic games or generalize earlier games or answer purposes of aesthetic enjoyment or play.

As before, the account of truth comes in two parts, one part that directly describes what practitioners do to achieve success, and a second part that offers a pragmatic validation of their practice. This time, however, more has been said to identify the sources of worth (or of aptness). Although the account of the pragmatic virtues is incomplete, (7) points in the directions in which it should be developed.[14]

Turn now to what appears to be an entirely unrelated realm. Many people have a religious account of ethical truth. They suppose that a statement of form "*A* is right" (where '*A*' refers to some action) is true just in case God commands *A*. Divine command accounts of ethical truth have been in trouble ever since Socrates raised embarrassing questions for Euthyphro, but they remain popular because of the apparent implausibility of the alternatives. Proposals that ethical statements are true when they correspond to the Good, or that rightness is a property of actions, remain unconvincing in the absence of any account of how to detect the Good or the alleged property of rightness. Since the available accounts seem to face

---

[13] Bellah's emphasis on the importance of play in the development of religious ritual is illuminating. Play also strikes me as important in the evolution of other domains of human practice, where it might initially seem irrelevant. Mathematics is a case in point.

[14] For more detail, see Kitcher 2012, ch. 7.

insuperable obstacles (and are usually formulated in technical language), the divine command theory survives, despite all its inadequacies.[15]

Here's a better way to think about ethical truth. Ethics is a human project, begun by our remote ancestors. Although they could not have conceived of what they were starting in this way, their introduction of patterns for their shared social lives was an attempt to solve a fundamental problem of the human situation. We have enough responsiveness to others to live together (and to want to live together), but too little to live together well. Ethics began as a form of social technology for overcoming the limits of our mutual responsiveness. For at least 50,000 years, and probably for significantly longer, groups of human beings have been working out the terms of their lives together, and the most successful experiments have endured in the ethical codes we have inherited. Although ethical progress has been rare, and typically achieved blindly, there's a coherent notion of ethical progress that can be applied to particular transitions in the history of ethical practice – for example in the abolition of chattel slavery. That notion of progress is grounded in problem-solving. Progress is *progress from*. As in the case of medicine, you make ethical progress by overcoming (or partially overcoming) the problems of the current situation. Ethical truth is what you get when you continue to make progress. As William James puts it, "Truth happens to an idea."[16]

The account just sketched can be made more precise. Start with the notion of an ethical code, conceived as a sequence of precepts demanding some actions, prohibiting others, and permitting a third class. Focus on demands, statements of form "Do *A*!" The statement "*A* is right" is the *descriptive counterpart* of the demand "Do *A*!" A demand is *progressively introduced* into a code just in case that demand partially solves some *well-grounded ethical problem*. A well-grounded ethical problem is either the original problem to which the ethical project responded – the problem of limited human responsiveness – or a problem generated from that original problem through the sequence of partial problem-solutions that have led to the code. A sequence of modifications of a code is *completely progressive* just in case all the elements changed are progressively introduced. Now for the account of ethical truth.

> (8) "*A* is right" is weakly ethically true just in case there's a code that has progressively introduced the demand "Do *A*!"

[15] This brusque dismissal is defended in Kitcher 2011, ch. 5.
[16] James 1975: Preface.

(9) "*A* is right" is strongly ethically true just in case it is weakly true and also the demand would be retained in any indefinitely long sequence of completely progressive modifications of the code.

Strong ethical truth is the important notion, capturing what is, I hope, an attractive idea: the ethical truths are those parts of ethical practice that constitute stably enduring elements to the problems that must be addressed by the ethical project.[17]

I now want to make what may appear to be an absurd claim: *strong ethical truth is fundamental to all species of truth*. That claim rests on two suppositions: first is the thought that all truth has a bipartite structure, including a pragmatic condition; second is the thesis that the pragmatic condition tacitly makes reference to strong ethical truth.

The need for the pragmatic condition is wittily exposed by Lewis Carroll in Alice's encounter with Humpty Dumpty, who declares that words shall mean whatever he chooses them to mean. There's a large infinity of potential languages that might be used in describing the world, playing symbolic games, or prescribing conduct, but the overwhelming majority of them are Humpty Dumpty languages, of no use or significance in any human project. The weak notion of truth is that of "truth-in-language-L," where no constraints debar the Humpty Dumpty languages. To talk of truth (period) is to discuss truth in some language worth employing.

To say that a language is worth employing is to recognize it as apt for the realization of some human purpose, and that is not only to relate it to what people contingently wish to do but to connect it with what is worth doing. Ultimately the various concepts of aptness that figure in the pragmatic conditions of the various species of truth are based in the projects that are worth pursuing, and in a notion of the good human life that is grounded in the progressive evolution of the ethical project.[18] Yet, for reasons that will become apparent in the next section, it's also worth having the weak conception of ethical truth, using it to explain the status of those projects and purposes that cannot claim the status of permanent worth, but are apt in particular cultural contexts or at historical stages.

To spell this out, let's say that aptness depends on answering to a valid human purpose. The validity of a purpose is often grounded in the fact that achieving the goal will contribute to some more fundamental

---

[17] For much more detail on the approach I have outlined here, see Kitcher 2011. I don't want to claim, however, that the treatment given there constitutes the complete development of the guiding ideas.

[18] The evolution of our ethical practices is a process that has proliferated unanticipated ways of human fulfillment. Again, see Kitcher 2011.

endeavor. Behind each valid purpose stands a chain of purposes, each more fundamental than its predecessor, that culminates in an ultimate purpose. I propose:

> (10) A purpose is strongly (weakly) valid just in case the statement that its associated ultimate purpose is worthwhile is strongly (weakly) ethically true.

(10) generates the notion(s) of aptness that should figure in the pragmatic conditions on any species of truth.

V

At last we can return to the original task of elaborating what I take to be the important insights of the tradition in which Bellah's work stands. Earlier, I introduced notions of fictional truth ((4) and (5)), and those might be extended directly to notions of mythical and religious truth. The discussion of the last section, however, especially the idea of the permanent progressiveness underlying strong ethical truth, suggests introducing more stringent demands on a notion of mythical truth. The weakest grade of mythical truth might be that achieved by those bodies of lore that address the valid purposes of people in a particular situation. The stronger grade of mythical truth requires "universal fruitfulness," in the sense that the myths speak to the purposes of all those whose cultures progress beyond the point at which the myths were progressively introduced. There are the locally valuable myths and the enduring myths.

The detailed studies offered by Durkheim, Bellah, and others provide many examples of the purposes served by religious rites, religious institutions, and religious myths. They also defend claims that those purposes are valid – at least weakly valid. It's possible to use the individual discussions as the basis for an account of aptness analogous to that I offered in the case of mathematical truth. Religious practices would be seen as apt in virtue of their release of tendencies to joyful play, to the cultivation of ethical sensibility, and to social solidarity. Suppose this provides a start on a concept of *religious aptness*. We could then count a transition as *religiously progressive* just in case it increased the aptness of the religious practice of the community. This would yield a notion of religious truth.

> (11) S is weakly religiously true just in case there's a community with a religious practice R and an extension of that practice R^*, such that the

extension involves the affirmation of S and the transition from R to R^* would be religiously progressive.

(12) S is strongly religiously true just in case S is weakly religiously true and the practice of affirming S would be retained in any indefinite sequence of religiously progressive modifications of R^*.

(11) and (12) are simply precise (pedantic?) ways of saying that the weak religious truths are those whose affirmation in religious practice helps fulfill the functions of religion (of which thinkers from Durkheim to Bellah have given us a partial account), and the strong religious truths are the enduring weak religious truths, the ones that, once introduced, are never abandoned (so long as the religious tradition makes progress).

Religious truth includes mythical truth, but isn't exhausted by it. There are many dimensions of religious practice, and statements play other roles besides the telling of stories or the recounting of lore. Given, however, that a religious practice does have a dimension in which lore is enunciated, and given that this dimension enhances the aptness of the practice, mythical truths would be covered as a special case. Mythical statements are those affirmed in this part of the practice, and they count as true (weakly or strongly) insofar as they satisfy the conditions on religious truth ((11) or (12)).

So far, this has been an attempt to clear Bellah's suggestive proposals of the charge that they are completely confused, and to provide a better clarification of the underlying ideas than that available within the (potentially dangerous) framework he draws from Schutz. Yet it is clear that my reconstruction introduces a suspect concept: many people are uneasy at talk of cultural progress. In the ethical case, there are occasional transitions in ethical practice – the abolition of slavery, the enhancement of opportunities for women, the rejection of prejudices against same-sex love – where a concept of progress is hard to resist, and I have argued that that concept can be coherently developed. Some transitions in religious practice are ethically progressive (as when previously stigmatized people are included – think of the parable of the Good Samaritan). But, given the diverse purposes religion serves, there are other modes of religious progress, some of them visible in liberation movements, others in the refinements of rites. The concept of religious progress applies to some well-known historical transitions – perhaps especially strikingly in the emergence of the five major religions of the axial age. Once the functions of religious practice are recognized, the concept can be demystified by conceiving progress in terms of refinement of functions. To repeat: comparing large traditions with respect to progressiveness may be impossible, but that doesn't mean we can't say, of a particular change within a religious practice, that it is progressive. A local notion of progress is nothing to be frightened of.

I want to bring out one particular type of religious progress, in order to reinforce my diagnosis of the dangers in Bellah's Schutzian framework. Religious practices make *purifying progress* when their adherents no longer confuse the types of truth pertinent to the religious affirmations they make. One prime instance of confusion is, of course, the affirmation of religious truths – mythical truths – as matters of historical fact. Another is to treat prayer as a conversation with a remote being. Religions make purifying progress when they amend their practices so as to instill in their followers psychological dispositions to avoid thinking of religious truths as true in the correspondence sense (that is, in accord with (3)). The ideal purifying progressive religions would be so explicit about this boundary that the mistake would be completely avoided. Whether there are any such religions is doubtful – and I am sure that those who engage in Schutzian talk of "multiple realities" will not achieve them.

The most sophisticated of contemporary atheists – Daniel Dennett, for example – think of purifying progress as the principal (if not unique) mode of religious progress. To the extent that they appreciate the Bellah–Durkheim points about the rich variety of human purposes religions have served, they take these purposes to be insignificant in comparison with the deceptions and self-deceptions that arise from inadequate boundaries among types of truth.[19] Purifying progress must come first. Any functions that remain to be served can be attended to once the ground has been cleared.

Because I view the Durkheim–Bellah tradition as recognizing important features of religious life, the insistence on purifying progress before everything else strikes me as insensitive to most versions of the human predicament. Yet there is something clearly correct about Dennett's demands, something that often slips from view in the writings of those who explore the richness of religious traditions. The notions of 'reality' and 'truth' are left so obscure that the treatment seems antithetical to purifying progress: it almost encourages the breach of boundaries, the gliding from one species of truth to another. Discussions of religion need a framework that allows for appreciating the many sides of religious practice and the valid functions served *and* the importance of purifying progress, a framework that accommodates Dennett's insights *as well as* Bellah's. Providing that framework has been the goal of this chapter.

---

[19] I suspect that Dennett would strongly dislike this formulation, viewing it as already fostering the kinds of confusions he is so anxious to avoid.

VI

I'll end with a brief look at an important debate. Secular humanists (like myself) are often accused of telling a simple, and historically false, story about the growth of the secular perspective, a "subtraction narrative" as it's called.[20] There's clearly a position – held by people less subtle than Dennett – that merits the charge: a position taking purifying progress to be the only type of progress achievable with respect to religion. Because I suppose that religious practices respond to important and pervasive features of human life, I recognize modes of religious progress in which religious traditions improve their efforts at satisfying the needs of their adherents. Were the clear recognition of the distinction between correspondence truth and religious truth to spread too rapidly, without available cultural institutions to take over the functions religions have served, the human losses would probably be far greater than the gains. Religions may also make progress through finding more effective ways of reinforcing important values[21] – and Ruse's continued sympathy for the Quakerism of his boyhood may stem from recognizing the Quakers as making just this sort of progress within the Christian tradition. So, in the secular humanism I favor (akin, I hope, to Ruse's own), it is seen as a positive doctrine, one that attempts to provide surrogates for the many dimensions of religious practice.

Insofar as it's possible to think of a broadly progressive future, it isn't one in which religion disappears, but one in which it metamorphoses into something else. Perhaps the myths that descend from the axial age will endure, perhaps not. There will be some collection of stories that continue to play the valuable role, but they may have been composed by quite different types of people – by poets and dramatists and novelists and film-makers and essayists and philosophers. There will be social institutions to develop the senses of identity and community that the traditional religions have fostered, but they may be entirely detached from the myths of the religions we have. Ancient texts may still be read by a few, but their importance will be viewed as stemming from their articulation of some ethical truth – and that ethical truth will be appreciated in its own right, as a truth of the sort characterized in (8) and (9). Everybody will be clear that the statements in the old myths are not true in the correspondence sense.

---

[20] See Taylor 2007; Calhoun 2011.
[21] This is a point articulated by Dewey in the closing pages of his *A Common Faith* (1934).

That vision of the future is not ungrounded speculation – for parts of it are already visible in the present. The route to the fully secular world-view of the future runs through the post-Enlightenment past, but it's one of complex change, not of mere "subtraction." The cultures that traverse that route accept an ever broader body of truths – not just factual truths (correspondence truths) but ethical truths and fictional truths as well. Myth becomes embedded in a broader category of fictions that contribute to the purposes religions have served. But the valuable stories may be constantly changing.

From this perspective, there are weak religious truths, statements whose affirmations have played a valuable role in particular cultures. The predicaments to which such parts of religious practice have responded will not disappear from the human future: people will continue to need ways of forging identities and achieving community, and these will be facilitated through their contemplation of bodies of lore not true in any correspondence sense; fictional truths will continue to be important to human lives. But I am skeptical about whether any particular fiction, even the myths of the axial age, is so deep and fundamental that it delivers strong religious truths. The only strong religious truths may well be ethical truths, and they are probably best explicitly regarded in this way. For secular humanists like me, illuminating fiction comes from different places. Shakespeare and Tolstoy, Joyce and Proust and Mann turn out to be more significant than any religious myths.

Perhaps, then, every myth is pertinent only in a range of cultural contexts, so that (apart from the ethical insights) there are no strong religious truths. If that is so, the only difference between the purified religious perspective (exemplified by sophisticated thinkers in the Durkheim–Bellah tradition) and secular humanism (of the sort I hope Ruse and I share) is a disagreement about the extent to which religious myths can be superseded by other fictions. And that is not a disagreement that should spark any serious quarrel.

REFERENCES

Bellah, Robert (2012) *Religion in Human Evolution*. Cambridge, MA: Harvard University Press.

Benacerraf, Paul (1973) "Mathematical Truth." *Journal of Philosophy* 70: 661–80.

Calhoun, Craig (2011) "Secularism, Citizenship, and the Public Sphere." In Craig Calhoun, Mark Juergensmeyer, and Jonathan van Antwerpen (eds.), *Rethinking Secularism*. New York: Oxford University Press, pp. 75–91.

Carroll, Lewis (2010) *Through the Looking Glass*. London: Penguin Books (first published 1871).

Chaisson, Eric (2001) *Cosmic Evolution*. Cambridge, MA: Harvard University Press.

Dewey, John (1934) *A Common Faith*. New Haven, CT: Yale University Press.

Dummett, Michael (1978) *Truth and Other Enigmas*. Cambridge, MA: Harvard University Press.

Durkheim, Émil (2008) *The Elementary Forms of Religious Life*. Mineola, NY: Dover.

Evans-Pritchard, E. E. (1956) *Nuer Religion*. Oxford: Clarendon Press.

Goodman, Nelson (1978) *Ways of Worldmaking*. Indianapolis: Hackett.

James, William (1975) *The Meaning of Truth*. Cambridge, MA: Harvard University Press.

(1981) *Principles of Psychology*. Cambridge, MA: Harvard University Press.

Kitcher, Philip (2001) *Science, Truth, and Democracy*. New York: Oxford University Press.

(2011) *The Ethical Project*. Cambridge, MA: Harvard University Press.

(2012) *Preludes to Pragmatism*. New York: Oxford University Press.

Kuhn, Thomas (1962) *The Structure of Scientific Revolutions*. University of Chicago Press.

Lear, Jonathan (1977) "Sets and Semantics." *Journal of Philosophy* 74: 86–102.

Taylor, Charles (2007) *A Secular Age*. Cambridge, MA: Harvard University Press.

(2011) "Western Secularity." In Craig Calhoun, Mark Juergensmeyer, and Jonathan van Antwerpen (eds.), *Rethinking Secularism*. New York: Oxford University Press, pp. 31–53.

PART II

Taxonomy and systematics

Consilience, historicity, and the species problem

Marc Ereshefsky

Introduction

The species problem is one of the big problems in biology and the philosophy of biology. For hundreds of years, biologists and philosophers have tried to answer the question: what is the proper definition of 'species'? And despite hundreds of years of work on this problem, there is still widespread disagreement over the correct answer. Michael Ruse, of course, has tackled the species problem (see Ruse 1969, 1971, 1973, 1987, 1988). (I say 'of course' because Ruse has written on every significant issue in the philosophy of biology.) Ruse's arguments concerning species are cogent and innovative. And they are frequently rehearsed by other philosophers 40 and 25 years after he introduced them.

Ruse's work on species addresses two philosophical issues. One is the ontological status of species: are species natural kinds akin to elements on the periodic table or are species individuals akin to particular organisms? The traditional and most popular view among philosophers is that species are natural kinds. In the 1970s, Ghiselin (1974) and Hull (1978) challenged that traditional view. Their species-are-individuals thesis is now the received view in the philosophy of biology. Not soon after Ghiselin and Hull introduced the species-are-individuals thesis, Ruse offered a rigorous defense of the view that species are natural kinds.

The other philosophical issue concerning species that Ruse has tackled is whether 'species' refers to a real category in nature or whether the species category is merely an artifact of our theorizing. This is an old question, predating Darwin. Ruse offers an innovative argument in favor of species realism —the view that the term 'species' refers to a real category in nature. To make his case, Ruse (1994) turns to his favorite philosopher, William Whewell, and he employs Whewell's consilience of inductions. Ruse's argument for species realism has recently been updated by Richards (2010).

Though Ruse's arguments concerning species are cogent and innovative, I will contend that they are flawed. Nevertheless, they are important
arguments, and numerous philosophers of biology still employ them. The
tenacity of Ruse's arguments testifies to their significance. Though much
of this chapter will be a critique of those arguments, I will offer a positive
answer to the species problem. In particular, I will suggest that when Ruse
and others argue against the species-are-individual thesis, they miss what
is most important about that thesis: that species are historical entities. I
will also try to clarify what it means to say that species are historical entities by developing the idea that species are path-dependent entities. When
it comes to the question of whether 'species' refers to a real category in
nature, I will offer a pragmatic form of species anti-realism. Such anti-
realism holds that the species category is not a natural category, yet the
word 'species' should not be relegated to the dust heap of outdated theoretical terms.

Historicity and species

Ruse's arguments concerning the ontological status of species are largely
a reaction to Hull's (1978) arguments on the topic. So let us start with
Hull's distinction between kinds and individuals and Hull's argument for
the species-are-individuals thesis. According to Hull, kinds are groups of
entities that function in scientific laws. Hull maintains that such laws are
true at any time and any place in the universe. Copper is a kind because
the law 'All copper conducts electricity' is true here and now as well as a
million years from now on some distant planet. In other words, an entity
is a member of the kind copper as long as it has certain theoretical properties. The parts of an individual, on the other hand, cannot be scattered
across time and space. They must exist in a particular space-time region.
Consider a paradigmatic individual, the dog Lassie. Certain dog parts are
only parts of Lassie if they are appropriately spatiotemporally connected.
Lassie parts, when they are parts of Lassie, cannot be scattered anywhere in
the universe. The same is true of more controversial individuals, according
to Hull, such as countries. Though Hawaii is not geographically contigu-
ous with any other part of the United States, that country is an individual
because its parts must occur within a restricted space-time region to be
parts of a single country.[1]

---

[1] Boyd (1999), Okasha (2002), and LaPorte (2004) reject the distinction between individuals and
  kinds arguing that the distinction is merely 'syntactic'. Though there are problems with Hull's formulation of the distinction, for example, his characterization of scientific laws, I think it is wrong to

Given this distinction between kinds and individuals, why does Hull think that species are individuals? His argument starts with the assumption that 'species' is a theoretical term in evolutionary biology. Hull (1978) argues that species are units of evolution in evolutionary biology, meaning that species are groups of organisms that evolve as a unit. Natural selection is the primary force that causes species to evolve. One way that selection causes a species to evolve is by causing a rare trait to become prominent within a species. For such evolution to occur, a trait must be passed down through the generations of a species. That requires that the organisms of a species are connected by reproductive relations: namely, sexual relations between parents (in sexual species), and parent–offspring relations between parents and offspring. Such relations require that organisms, or their parts (gametes and DNA), come into contact. Consequently, evolution by selection requires that the generations of a species are spatiotemporally connected. In other words, the organisms of a species cannot be scattered throughout the universe but must occupy a particular space-time region. Given that species are units of evolution, they are individuals and not kinds.

With the difference between kinds and individuals and Hull's argument for species being individuals in hand, we can turn to Ruse's rebuttal of the individuality thesis. Ruse offers several arguments against species being individuals. Let us go through those arguments. Along the way we will get to the crux of the individuality thesis: that species are historical entities.

Ruse's (1987, 232–34; 1988, 56) first argument against species being individuals involves the units of selection controversy. In a nutshell, Ruse's argument runs like this: individuals are units of selection. The majority of biologists that work on natural selection doubt that species are units of selection (they think that organisms are the units of selection). Therefore, we should doubt that species are individuals. In his words: "What some Darwinians find particularly troublesome about the species-as-individuals thesis is that it seems to flatly go against the renewed biological emphasis on individual selection" (Ruse 1988, 56). I do not want to wade into the debate over the units of selection, but merely show that Ruse is wrong to think that the units of selection debate sheds light on the ontological status of species.

Hull does not offer one account of biological individuality but several. He offers his basic notion of individuality in his work on species

reject the difference between individuals and kinds because to do so inappropriately conflates two distinct ways scientists construct classifications (Ereshefsky 2010a). This debate, however, can be put to one side because Ruse (1987, 1988) adopts Hull's dichotomy.

(Hull 1978): individuals must be spatiotemporally restricted entities. Hull also offers a twofold account of individuality – a refinement on his basic notion – in his work on natural selection (Hull 1980). According to Hull, two different kinds of individuals are required for natural selection to occur: replicators and interactors. Replicators and interactors must satisfy his basic criterion of individuality – they must be spatiotemporally restricted entities. In addition, replicators and interactors have their own specific criteria. For Hull, when we ask whether an entity is an individual we need to ask whether it is an individual of a certain type: is a species an individual *qua* evolutionary unit or *qua* unit of selection? Hull argues that as evolutionary units species must be individuals. He is not arguing that they are individuals in selection. Indeed, Hull (1980, 324, 327) clearly doubts that species are units of selection. Thus, Ruse's first argument against the species-are-individuals thesis is misplaced: he needs to show that as units of evolution species need not be individuals.

Ruse's (1987, 234–35) second argument turns on the question of whether species are sufficiently integrated by gene flow to be individuals. Ruse suggests that gene flow provides "the kind of integration required for individuality" (1987, 234). He points out that many species are not integrated by gene flow. He concludes that many species are not individuals. The success of this argument turns on the question of whether the presence of gene flow among the populations of a species is necessary for a species to be an individual.

Hull (1978, 343–44) suggests that three processes, along with genealogy, can cause species to be distinct evolution units. One is gene flow among the members of a species. The transmission of genes among the organisms of a species through interbreeding can cause those organisms to evolve as a unit. Hull also suggests that genetic homeostasis and selection can cause unity among the members of a species. Following Eldredge and Gould (1972) and Mayr (1970), Hull argues that when organisms of a species share similar homeostatic genotypes those organisms remain similar despite their occurring in different environments and being exposed to different mutations. Following Ehrlich and Raven's (1969) seminal work on stabilizing selection, Hull suggests that selection can cause the members of the species to evolve as a unit.

Returning to Ruse's argument, Ruse is correct that many species lack the requisite gene flow that would cause them to be evolutionary units. Many species of sexual organisms consist of geographically isolated populations. Yet despite insufficient gene flow among their populations, they are unitary species. More pressing is the fact that most of life on this planet reproduces asexually not sexually. Gene flow only occurs when sexual organisms interbreed. There is no interbreeding among asexual organisms. Furthermore, it is a well-known fact that most of life on this planet is microbial, and the vast majority of microbes do not produce sexually (Ereshefsky 2010b). So, yes, Ruse is correct that many species are not integrated by gene flow. Does that, then, show that most species are not individuals? Recall that Ruse writes that gene flow provides "the kind of integration required for individuality" (1987, 234). However, other processes besides gene flow, namely selection and genetic homeostasis, provide such integration.

Ruse's emphasis on gene flow misses the heart of the species-are-individuals thesis, namely that species are genealogical entities. Species must be genealogical entities and that is sufficient to make them individuals. Recall Hull's evolutionary unit argument cited earlier. Species are first and foremost units of evolution. That requires that the different generations of a species are connected by parent–offspring relations. Otherwise, the changes caused by various evolutionary forces will not be passed down from generation to generation. That is why, according to Hull, species must be individuals, where being an individual merely means being a spatiotemporally continuous (and hence restricted) entity. The heart of the species-are-individuals thesis has nothing to do with the existence of gene flow within a species. It is about species being evolutionary units. The passing on of genes from parent to offspring (genealogy) is required. A casually integrating force like gene flow is not required, because there are other processes besides gene flow that cause species unity.

Let us turn to Ruse's strongest argument against species being individuals. Recall that one of the main tenets of the species-are-individual thesis is that species are spatiotemporally continuous entities. The generations of a species must be genealogically connected if a species is to be a unit of evolution. Or to put it in negative terms, a species cannot consist of genealogically disconnected populations. Ruse argues that this central tenet of the species-are-individuals thesis is wrong. In Ruse (1988, 56), he writes: "Suppose a new organism is produced through polyploidy. Suppose then that all members of this new species are destroyed, and then at some later point new, similar organisms are produced. Surely we have new members of the same species, not a new species?" Polyploids have a different number of chromosomes from organisms in their parental species. As a result, they cannot interbreed with members of their parental species. Sometimes polyploidy culminates in speciation, but often it does not (Briggs and Walters 1984, 242). This is an important point: polyploidy does not automatically cause the existence of a new species; it is just the potential start of a new species. Whether speciation occurs depends on whether the new polyploids and their descendants flourish. Thus, Ruse's hypothetical example of a genealogically disconnected species – one with two origins – is biologically questionable: the mere occurrence of polyploidy is not a speciation event. (We will return to the case of polyploidy shortly.)

Ruse 1987 offers a different example to motivate the plausibility of a species having multiple origins.

> Today, through recombinant DNA techniques and the like, biologists are rushing to make new life forms. Significantly, for commercial reasons the scientists and their sponsors are busy applying for patents protecting the new creations. Were the origins of organisms things which uniquely separate and distinguish them, such protections would hardly be necessary. Old life form and new life form would necessarily be distinct. Since apparently they are not, this suggests that origins do not have the status claimed by the [species-are-individuals] boosters. (Ruse 1987, 235–36)

An odd thing about this argument is that it assumes that commercial interests in biotechnology are decisive in the debate over the ontological status of species. Yet parties in this debate generally see the debate decided by scientific theory. Those worried about genetic patents are not obviously concerned about whether they have created a new species *qua* evolutionary theory. I read the commercial interests surrounding such patents as not about species-hood but the patenting specific genotypes. Furthermore, prominent theoretical definitions of the term 'species' (what biologists call 'species concepts') do not define species in terms of specific genotypes. Mayr's (1970) Biological Species Concept defines a species as a group of interbreeding organisms reproductively isolated from other such groups. The various Phylogenetic Species Concepts (Baum and Donoghue 1995) define species as genealogical segments on the Tree of Life. Even Mallet's (1995) Genotypic Cluster Concept does not define a species by a single genotype. For Mallet, a species consists of a statistically defined cluster of similar but different genotypes. Ruse may be correct about the patenting of genotypes, but such commercial interests do not show that species are or should be defined by distinct genotypes.

Nevertheless, there is something appealing to a number of philosophers about the idea that species can have multiple origins. This suggestion is not only made by Ruse but a number of philosophers, including Kitcher (1984), Boyd (1999, 2010), Elder (2008), and Devitt (2008). There is, however, a fundamental aspect of species they are missing, namely that species are historical entities. Why should we think that are species historical enti-ties? The short answer is that species are path-dependent entities. In what follows, I will fill this out by first introducing the notion of path depend-ency and then explaining why species are path-dependent and hence his-torical entities.

Desjardins (2011) draws the following distinction between two types of historical entities. There are entities whose properties depend on ini-tial conditions, and there are entities whose properties depend on initial conditions and the historical path taken after those initial conditions. According to the first notion of historicity, the probability that an entity has a certain property is a function of initial conditions. For example, the probability that Joe will die from radiation poisoning is largely dependent on how much radiation Joe was exposed to during the Chernobyl atomic power plant disaster. According to the second notion of historicity – path dependency – not only do initial conditions affect the probability of an outcome, so do events along the path from initial conditions to the out-come, as well as perhaps the order of those events. Consider the case of Michigan State biologists producing 12 identically cloned *E. coli* popu-lations, and then placing them in identical but separate environments and letting them evolve for thousands of generations (Desjardins 2011). After about 10,000 generations, those populations evolved different adap-tive traits. According to the biologists involved, such variation was due to the organisms in different populations having different mutations. The biologists also argued that the mutations in the various populations came in different temporal orders, and mutation order was important because prior mutations created the genetic background for latter mutations to be adaptive. In other words, these populations started with identical genotypes and were placed in identical environments, yet because those populations had different mutations and different mutation orders, they acquired varying traits. The acquisition of those traits, in other words, was a path-dependent process.

Let us return to species. Species are path-dependent entities because speciation is a path-dependent process. To see why, consider the allo-patric model of speciation, the most widely accepted form of speciation among biologists. According to that model, speciation begins when a population is isolated from the main body of its parental species (Ridley 1993, 412). When applied to sexual species, allopatric speciation is con-sidered complete when a population is reproductively isolated from the members of the parental species: that is, organisms in the parental and new species cannot interbreed and produce fertile offspring. Such reproductive isolation occurs when organisms in parental and new species have isolating mechanisms that prevent them from interbreeding and producing fertile offspring. Those mechanisms may be pre-zygotic mechanisms that prevent interbreeding, such as incompatible sexual physiology; or they might be post-zygotic mechanisms that prevent offspring from being viable or fertile. How do such isolating mechanisms arise? According to Mayr (1970, 327), isolating mechanisms are by-products of new adaptations in new species. For example, Podos (2001) argues that some of Darwin's finches are reproductively isolated because they have different mating calls. Furthermore, their having different mating calls is a by-product of evolution for specialized beaks for eating different foods. Some beaks are long and good for probing in wood, others are short and can gather seeds on the ground. Now ask, what is the common source of new adaptations? Answer: mutations and previous changes in the genetic background of an organism that allow a new mutation to be beneficial. Here, then, is the point. Mutations and mutation order are important causes of speciation. Different populations have different mutations and mutation order (as well as differences in the effects of genetic drift) even if those two populations start with identical clones and identical environments. The upshot is that speciation is a path-dependent process: vary the path and it is very, very unlikely the same species will be produced. I should add that it is not empirically impossible. The point here is that given what we know about evolution, it is very unlikely.

Let us go back to Ruse's polyploidy example. Suppose, hypothetically, there are two populations of organisms that are the result of separate polyploidy events. Coincidentally, the two populations start with organisms with identical chromosomes. Furthermore, both populations are reproductively isolated from their common parental species. Should we then say there is a new species even though it consists of two genealogically disconnected populations? As mentioned earlier, the answer is no. Here is where path dependency comes in. For a new population to become successful and become a new species, it needs to be able to exploit a niche different than the niche occupied by its parental species. How does a new population acquire the ability to exploit a new niche? Some adaptive difference must arise among those organisms through mutations and changes in their genotypes. As we have seen, organisms in different populations are exposed to different mutations and in different mutation orders. Path dependency is crucial in the completion of speciation, and even initially identical polyploids undergo different paths.

Stepping back from these details, we see that Ruse's arguments that species may be spatiotemporally discontinuous entities – that they may not be individuals – face two challenges. First, there is Hull's evolutionary unit argument, that species are entities that evolve via selection, and selection requires the different generations of a species to be genealogically connected. Second, species are path-dependent entities because speciation is a path-dependent process. That two populations consist of identical clones is insufficient to make those populations parts of one species. Whether there is a new species depends on later events in speciation, and it is very unlikely that two isolated populations will undergo the same path of events. It is possible, but unlikely according to current biological theory.

Consilience and species

Let us change gear and turn to Ruse's contribution to the other big philosophical question concerning species, namely whether the term 'species' refers to a natural category or is merely an artifact of our theorizing. His answer to this question is innovative and significant. In determining whether species is a natural category, Ruse (1994) turns to his favorite philosopher, William Whewell. Ruse believes that Whewell's consilience of inductions is a good indicator of a concept's naturalness. He applies it to 'species' and argues that because 'species' satisfies the consilience of induction we have good reason to believe that species is a real category (Ruse 1987, 1988). In what follows, I will not question whether Whewell's consilience of inductions is a good method for evaluating whether a concept corresponds to a natural category. Instead, I will question whether that method applies to 'species'.

According to Whewell (1968, 138–39), the consilience of inductions "takes place when an Induction, obtained from one class of facts, coincides with an Induction, obtained from another different class. This Consilience is a test of the truth of the Theory in which it occurs." For example, evidence from terrestrial phenomena, such as the movement of balls and pendulums, confirms Newton's laws, and so does evidence from celestial phenomena, such as the movement of Earth's Moon and the rotation of the planets around the Sun. Together these different classes of facts provide a consilience of inductions for Newtonian mechanics. Ruse observes that Whewell applies the same general principle to classification: "The Maxim by which all Systems professing to be natural must be tested is this: – *the arrangement obtained from one set of characters coincides with the arrangement obtained from another set*" (Whewell 1840, vol. I, 521;

quoted in Ruse 1987, 238). Or as Ruse (1987, 238) describes it: "[a] natural classification is one where different methods yield the same result."

Ruse applies the consilience of inductions to the species problem by considering the different ways that biologists construct classifications of species. He argues that those different ways of constructing classifications coincide:

> Coming back to organic species, we see that we have a paradigm for a natural classification. There are different ways of breaking organisms into groups, and they *coincide*! The genetic species is the morphological species is the reproductively isolated species is the group with common ancestors. (Ruse 1987, 237; also see 1969, 111–12 and 1988, 54–55)

By 'morphological species' he means "groups of similar looking organisms, with gaps between the groups" (Ruse 1987, 226). Reproductively isolated species are groups of organisms that satisfy Mayr's (1970) Biological Species Concept. Genetic species are "overall *genetic* similarity clusterings, being separated from other such gaps" (Ruse 1987, 227). For groups with common ancestors, he refers to Simpson's (1961) Evolutionary Species Concept: a "species is a lineage ... evolving separately from others and with its own unitary evolutionary role and tendencies" (quoted in Ruse 1987, 227).

It would be wonderful if these different types of groups did coincide, but they do not. Consider classifications based on overall morphological similarity and those based on interbreeding. The fruit flies *Drosophila persimilis* and *Drosophila pseudoobscura* are almost morphologically identical but are reproductively isolated from one another (Mayr 1982, 281). Alternatively, consider genetic species and reproductively isolated species. In some cases of flies, fish, and frogs there is more genetic variability within an interbreeding species than between two reproductively isolated species (Ferguson 2002). One might respond that such cases are the exception and generally the different approaches to species do line up. But that is not the case. The discrepancies among modern approaches to species are widespread. Mayr's Biological Species Concept and the Phylogenetic Species Concept (which comes in various versions; see Baum and Donoghue 1995) are the most popular approaches to species among biologists. Yet they carve the organic world in different ways. For cladists, all taxa are monophyletic: they include all and only the descendants of a unique ancestor. Unique ancestry is the key. Cladists identify taxa as branches on the Tree of Life, and species are the smallest twigs on that tree. Those that support the interbreeding approach identify groups of interbreeding sexual organisms.

They want to identify distinct gene pools: pools of shared genes. Both the phylogenetic and interbreeding approaches to species highlight significant aspects of evolution: genealogical lineages and gene pools. Yet many cladistic lineages are not groups of interbreeding organisms, and many groups of interbreeding organisms are not cladistic lineages.

Consider cases of the first sort. Only sexual organisms reproduce by interbreeding, so the interbreeding approach to species only applies to sexual organisms. Asexual organisms reproduce by a variety of other means, such as budding, binary fission, and vegetative reproduction. The interbreeding approach does not place such organisms into species. They are simply not classified into species. The phylogenetic approach does classify asexual organisms. All that matters for the phylogenetic approach is whether a group of asexual organisms is monophyletic. So a major discrepancy between the interbreeding and phylogenetic approaches is that the latter but not the former classifies asexual organisms into species. This is no small discrepancy, for most of life, whether it be the number of organisms on Earth or the percentage of Earth's biomass, is asexual (Hull 1988, 429; Templeton 1992, 164). Thus, for most of life the interbreeding and phylogenetic approaches do not coincide.

Another major discrepancy between the interbreeding and phylogenetic approaches concerns ancestral species. As we saw in the previous section, the most widely accepted model of speciation, allopatric speciation, holds that speciation starts when a population becomes isolated from the main body of a species. That isolated population undergoes a 'genetic revolution' and, if successful, becomes a new species. The parental species – the ancestral species – remains intact. The interbreeding approach allows the existence of ancestral species, but the phylogenetic approach does not. A figure can help show this (Figure 4.1). According to the interbreeding approach, when such speciation occurs, there are two species: C, which is the new species; and A+B, which is the ancestral species. The phylogenetic approach denies that there are two species in such cases. For the phylogenetic approach, a species must be monophyletic and contain all and only the descendants of a common ancestor. The ancestral species A+B is not monophyletic: some of A's descendants are not in A+B. So, on the phylogenetic approach, there are not two species present, but either one species or three species. If there is one species, it consists of A, B, and C. If there are three species, they are species A, which has gone extinct, and species B and species C. Either way, the interbreeding and phylogenetic approaches give different answers to the number of species present in such situations. This is no small discrepancy, because there are countless ancestral species

according to the interbreeding approach but none according to the phylogenetic approach.

Thus far, I have focused on the two most popular approaches to species
among biologists that study eukaryotes. Pretty much all of the philosophical discussion of species focuses on species concepts developed for eukaryotes. Yet most of life is microbial (Rosselló-Mora and Amann 2001, 40).
This is a serious lacuna in the philosophical literature concerning species
because microbiologists offer their own species concepts. Those concepts
also produce inconsistent classifications of organisms and further undermine the claim of consilience among species concepts.

One microbial species concept, the Recombination Species Concept,
asserts that species are groups of microbes whose genomes can recombine
(Dykuizen and Green 1991). The motivation is that though microbes generally do not reproduce sexually, they form gene pools of organisms connected by recombination.[2] Another microbial species concept is Cohan's
(2002, 467) ecological concept in which a "species in the bacterial world
may be understood as an evolutionary lineage bound by ecotype-periodic
selection." A third approach to microbial species uses genetic data to
determine phylogenetic relations (Stackebrandt 2006). Just as in the case
of eukaryote species concepts, these microbial concepts often classify the
same group of organisms into different species. For example, in the genus
Thermotoga the same group of organisms forms one species according
to the Recombination Species Concept but multiple ecological species
according to Cohan's ecological approach (Nesbø *et al.* 2006).

Then there is the phylogenetic approach to microbial species, according
to which the same group of organisms can be classified in multiple ways
depending on which genes are used. For example, Wertz *et al.* (2003) suggest using core genes to classify microbes into phylogenetic species. Core
genes control such functions as cell division and metabolism. It is assumed
that core genes are evolutionary stable because a change in them would
greatly affect the viability of an organism. The problem, however, is that
there are multiple core genes in a microbe. Wertz *et al.* offer a case where
six different core genes from the same genome are used, and the result is
six different phylogenetic trees. Besides core genes there are other types
of genes microbiologists use to construct classifications. Some biologists

---

[2] It is worth pointing out that the Recombination Species Concept is not a version of the Biological
Species Concept. Interbreeding species are (relatively) closed gene pools due to pre- and post-
zygotic mechanisms. There are no such mechanisms among the members of recombination species.
Moreover, there is frequent lateral gene flow among microbial species. As a result, interbreeding
species are (relatively) closed gene pools, whereas recombination species are open gene pools.

use 16S rRNA genes. Others use DNA:DNA hybridization and look for a reassociation value of 70 percent or higher. These two ways of identifying species also produce conflicting species classifications (Rosselló-Mora and Amann 2001, 47; Stackebrandt 2006, 35). One might ask whether a particular type of genetic data better captures microbial phylogeny than another. The answer is no. Different genes simply have different phylogenies even though they are parts of the same genome (Doolittle and Bapteste 2007). In other words, various gene phylogenies run through a group of organisms and place those organisms into a plurality of phylogenetic species.

Stepping back from these details, we see that the two major species approaches to eukaryotes, the interbreeding and phylogenetic approaches, often provide conflicting classifications. Furthermore, different approaches to microbial species often sort the same group of organisms into different species. Clearly, the concept of 'species' does not satisfy Whewell's consilience of inductions. Facts from biological taxonomy undermine Ruse's argument for the naturalness of the species category.

In his recent book, Richards (2010) concurs with this assessment of Ruse's argument:

> The problem with Ruse's proposal ... is that it does not look as if this consilience is really forthcoming in a direct and simple manner ... If there really were a developing consilience, then we would presumably not see the proliferation of species concepts that group organisms inconsistently (Richards 2010, 130).

Nevertheless, Richards believes that a revised version of Ruse's argument can be deployed. Richards suggests that "if we apply the consilience idea to the hierarchical models of Mayden and de Queiroz, the prospects are more promising. Ruse's analysis may be on the right track, *if* we take into account the division of conceptual labor" (Richards 2010, 130). Let us review Richards's revised consilience argument and see whether it can establish the naturalness of the species category.

Richards's argument relies on Mayden (2002) and de Queiroz's (2005, 2007) work on species. Mayden and de Queiroz recognize major discrepancies among prominent approaches to species, but they contend that there is an important commonality among them. All such approaches assume that species are "separately evolving metapopulation lineages" (de Queiroz 2005, 1263). De Queiroz calls this view of species the "General Lineage Concept." According to Mayden, this concept "serves as the logical and fundamental over-arching conceptualization of what scientists hope to discover in nature behaving as species. As such, this concept can be argued to serve as the primary concept of diversity" (2002, 191). How is the General Lineage Concept related to other approaches to species? According to de Queiroz, the properties that proponents of other approaches disagree on (for example, successful interbreeding and monophyly) are merely properties that serve as "evidence for inferring the boundaries and numbers of species" (2005, 1264). Proponents of prominent species concepts are confusing "methodological" disagreements with "conceptual" ones (de Queiroz 2005, 1267). Consequently, their disagreements are not really over the definition of 'species' but over evidential and operational issues.

We can now see why Richards calls Mayden and de Queiroz's approach to species 'hierarchical'. There is one primary approach to species: all species are genealogical lineages. All other approaches to species, such as the Interbreeding and Phylogenetic Species Concepts, are secondary approaches that highlight the different types of evidence used for identifying species. In Richards's (2010, 142) words, Mayden's and de Queiroz's approach is "theoretically monistic and operationally pluralistic." Theoretically all species are genealogical lineages. Operationally, different biologists use different types of evidence for recognizing such lineages.

How does Richards's updated consilience argument for the existence of the species category fare? First, note that Richards's argument is different from Ruse's. Ruse's argument focuses on the proposition that though biologists use different approaches to species, those approaches tend to classify a group of organisms the same way. Ruse's argument relies on the occurrence of classificatory consilience. That sort of consilience is not a part of Richards's argument. Richards readily admits that different approaches to species will often sort the same group of organisms into different classifications. Richards instead relies on theoretical consilience: though biologists classify organisms into conflicting classifications, they nevertheless agree that species are genealogical lineages.

Richards's theoretical consilience, I will suggest, fares no better than Ruse's classificatory consilience. In brief, the counterargument to Richards's argument is this: biologists do not think that all genealogical lineages are species; they hold that species are a particular type of genealogical lineage. Moreover, they disagree on which type of lineage constitutes a species. Consequently, there is no theoretical consilience concerning 'species'. Let me unpack this counterargument. I agree with Richards that biologists believe that species are genealogical lineages. However, biologists also think that other Linnaean taxa are genealogical lineages: subspecies are lineages, so are genera, families, and so on. Being a genealogical entity does not distinguish species from other types of lineages. Biologists believe that species are a particular kind of genealogical lineage, but they disagree on which kind of lineage. As we have seen, some biologists believe that species are lineages of interbreeding populations. Others think that species are monophyletic lineages. Still others think species are lineages of organisms exposed to common selection regimes (see van Valen 1976). Because biologists disagree over which kind of lineages form species, there is no theoretical consilience concerning 'species'.

One might respond that species are nevertheless genealogical lineages, so Richards has given the proper definition of 'species' and solved the species problem. However, the problem with Richards's answer is that being a genealogical lineage is merely a necessary property of species. Unless which type of lineage is specified, we have an approach that identifies all Linnaean taxa (species, genera, families, etc.) as species, and that certainly does not solve the species problem. We need to specify which lineages are species. But once we specify which type of lineage is a species lineage, then there is no theoretical consilience concerning 'species'.

Ruse's original idea of applying the consilience of induction to the species problem is an innovative one. What better way to show that a scientific concept is tracking a real category than the consilience of different approaches to that concept? Unfortunately, neither Ruse's classificatory consilience nor Richards's theoretical consilience is successful. The problem highlighted here is not with the consilience of inductions, but with its application to biological taxonomy. There is no consilience among theories of species, and there is no general consilience among classifications involving species. Our theoretical conception of species stubbornly resists unification.

This result not only applies to Ruse's consilience argument and Richards's updated version, but also to other recent attempts to unify the species category. For instance, Brigandt (2003) and Griffiths (2007) write about a particular type of phenomena they call "species phenomena." However, there is no single type of phenomena that biologists agree upon as species phenomena (Ereshefsky 2010b). For example, supporters of the interbreeding approach believe that only sexual organisms form species. Supporters of the phylogenetic approach believe that only monophyletic lineages form species. Then there is the contrast between sexual and asexual species, and the contrast between eukaryotic and prokaryotic species. Different approaches to species recognize different types of phenomena as species. Wilson *et al.* (2009) also try to unify the species category. They write that there are "causally basic features that most *species* share." All species taxa are indeed genealogical entities and have many processes in common (for example, their organisms reproduce and their genes mutate). But those features do not set species taxa apart from other types of taxa, such as subspecies and genera. As we have seen, biologists are sharply divided on which causal properties set species apart from other types of taxa: some say interbreeding, others say selection factors or developmental homeostasis, still others say all three. The different arguments for the naturalness of the species category vary in which aspect of species is claimed to unify the species category. There is Ruse's consilience of classifications versus Richards's consilience of theories. There is Brigandt's and Griffiths's focus on species phenomena versus Wilson *et al.*'s focus on species' causal processes. Despite philosophers' best efforts, the biological world is uncooperative when it comes to unifying the species category.

The species problem

Let us take stock and draw some general conclusions. Earlier we saw that Ruse suggests that species need not be historical entities. However, that assertion conflicts with biological theory. Species are genealogical entities that undergo path-dependent processes. Species are not simply groups of identical organisms with the same start-up conditions, as Ruse and others suggest. Speciation is a path-dependent process involving a number of generations, a number of events, and events in a particular order. It is unlikely, given what biological theory tells us, that a particular speciation process will repeat itself. Ruse also argues that the concept 'species' refers to a real category in nature. We have seen that his consilience argument and Richards's updated version both fail: the species category has neither classificatory nor theoretical unity.

These results seem to leave us in an awkward position: species are historical entities yet there is no species category in nature. I would like to dispel the idea that this conclusion is paradoxical or untenable. Consider the distinction between species taxa and the species category. Species taxa are those individual taxa we call 'species', such as *Homo sapiens* and *Canis familiaris*. The species category is a more inclusive entity. It contains all those taxa we call 'species'. The conclusions of this chapter suggest that the species category does not exist outside human taxonomic practices. However, that should not cast doubt on the existence of those lineages we call 'species'. That is, the species category may not exist, but the lineages *Homo sapiens* and *Canis familiaris* do. To put it slightly differently, we might agree that there is a Tree of Life. (Or a bush of life if horizontal gene transfer is extensive.) *Homo sapiens, Canis familiaris,* and other taxa that we call 'species' are parts of that tree. It just happens that the Linnaean grid of ranks (species, genus, and so on) we use to classify those taxa is fictitious.

One might go along with this conclusion but wonder why should we continue using the word 'species' if there is no species category in nature. In fact, some writers suggest that the ambiguity of 'species' should cause us to use alternative and more precise terms such as 'biospecies', 'phylospecies', and 'least inclusive taxonomic unit' (Grant 1981; Ereshefsky 1992; Pleijel and Rouse 2000). Others suggest getting rid of the word 'species' and see no need to find a replacement (Mishler 1999, 2003). The aim to achieve an unambiguous and precise scientific language may be a worthy ideal but it is an impractical one (Kitcher 1984), especially when it comes to 'species'. The word 'species' is firmly entrenched in scientific discourse. It occurs in biology textbooks, field guides, and systematic studies. It is also entrenched in non-scientific discourse, for example, in governmental laws. Eliminating 'species' from biology and elsewhere would be an arduous task.

More importantly, there is no pressing need to eliminate the word 'species'. Some worry that the ambiguity of 'species' will cause confusion in biology (Hull 1978; Baum 2009). There is a simple way to deal with this problem, and it is a method that biologists do use to avoid confusion over the word 'species'. If the meaning of 'species' affects the understanding of a biological study, then the author of that study should be clear about his or her use of 'species'. In a biodiversity study, for example, a biologist should say whether numbers of interbreeding lineages or numbers of phylogenetic lineages are being counted. As Marris (2007) points out, some biodiversity studies count the number of interbreeding lineages, while others count phylogenetic lineages. The problem is that when the numbers from these studies are compared, like is not being compared to like. Two different types of biodiversity are falsely assumed to be one type of biodiversity. Another reason we should be explicit about the approach to species being used is that knowing a lineage's type can help us preserve a lineage. If different types of lineages are bound by different processes, then we need to know which type of process is crucial for maintaining the lineage we are trying to preserve.

There are other situations in which stating a particular approach to species is unnecessary for understanding the case at hand. If we merely want to indicate that one taxon is more inclusive than another taxon, we can call the more inclusive taxon a 'genus' and the less inclusive taxon a 'species' without specifying the type of species in question. The hierarchical

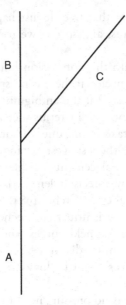

Figure 4.1 According to the interbreeding approach: A+B is a species and C is
a species. According to the phylogenetic approach: A, B, C are each subspecies;
or A, B, C are each species

relation between the two taxa is conveyed by 'species' and 'genus' without
saying whether the less inclusive taxon is an interbreeding or a phylogenetic lineage. Similarly, we can refer to a taxon as 'predator species' and
another as a 'prey species' and convey their relation without mentioning a
particular approach to species.

The answer to the species problem suggested here has three parts: (1)
doubt the existence of the species category; (2) do not doubt the existence of those taxa we call 'species'; (3) continue using the word 'species'.
Arguably, this approach to the species problem was how Darwin dealt
with the problem. What Darwin meant by 'species' and how he addressed
the species problem is highly controversial (Ghiselin 1969; Mayr 1982;
Beatty 1992; Stamos 2007; Mallet 2008; and Ereshefsky 2010c, 2011).
Some believe that Darwin was skeptical of the species category but not
those lineages called 'species' (Ghiselin 1969; Beatty 1992; and Ereshefsky
2010c, 2011). That raises the question: if Darwin was skeptical of the species category, why did he continue using the word 'species' throughout
his writings? According to Ghiselin (1969) and Beatty (1992), Darwin
kept using the word 'species' for practical reasons. They argue that

Darwin's primary objective in the *Origin of Species* was to convince biologists of his theory of natural selection. Attempting to reform language would get in the way of that aim. Consequently, Darwin kept using 'species' but denied that it had any theoretical meaning. For Darwin, the word referred to those lineages called 'species' by competent naturalists ([1859] 1964, 47). With that strategy in hand, Darwin could communicate his theory to others by arguing that those lineages called 'species' are the result of natural selection, but at the same time he did not have to undertake the impractical task of telling biologists to stop using the word 'species'.

The evidence, I believe, points to Darwin being a species taxa realist yet a species category anti-realist. However, I do not think consensus among Darwin scholars over what Darwin truly thought about species will come soon. Darwin played his cards very close to his chest on this issue. The historical evidence may stubbornly leave this issue unresolved. I am, however, more optimistic about the species problem. Though there is still widespread disagreement on the solution to that problem, I believe significant progress had been made. Our knowledge of the role of 'species' in biological theory is richer. Furthermore, we have a better understanding of what a proper definition of 'species' should look like. Many have made positive contributions to our understanding of species, including Ruse. His philosophical arguments concerning the nature of species are among the best, and philosophers continue to rehearse versions of those arguments 25 and 40 years after Ruse introduced them.

REFERENCES

Baum, D. (2009) "Species as Ranked Taxa." *Systematic Biology* 58: 74–86.

Baum, D. and M. Donoghue (1995) "Choosing among Alternative 'Phylogenetic' Species Concepts." *Systematic Biology* 20: 560–73.

Beatty, J. (1992) "Speaking of Species: Darwin's Strategy." In M. Ereshefsky (ed.), *The Units of Evolution*. Cambridge, MA: MIT Press, pp. 227–46.

Boyd, R. (1999) "Homeostasis, Species, and Higher Taxa." In R. Wilson (ed.), *Species: New Interdisciplinary Essays*. Cambridge, MA: MIT Press, pp. 141–86.

(2010) "Homeostasis, Higher Taxa and Monophyly." *Philosophy of Science* 77: 686–701.

Brigandt, I. (2003) "Species Pluralism Does Not Imply Species Eliminativism." *Philosophy of Science* 70: 1305–16.

Briggs, D. and S. Walters (1984) *Plant Variation and Evolution*. Cambridge University Press.

Cohan, F. (2002) "What Are Bacterial Species?" *Annual Review of Microbiology* 56: 457–87.

Darwin, C. ([1859] 1964) *On the Origin of Species: A Facsimile of the First Edition.* Cambridge, MA: Harvard University Press.

De Queiroz, K. (2005) "Different Species Problems and Their Resolution." *BioEssays* 27: 1263–69.

(2007) "Species Concepts and Species Delimitation." *Systematic Biology* 56: 879–66.

Desjardins, E. (2011) "Historicity and Experimental Evolution." *Biology and Philosophy* 26: 339–64.

Devitt, M. (2008) "Resurrecting Biological Essentialism." *Philosophy of Science* 75: 344–82.

Doolittle, W. F. and E. Bapteste (2007) "Pattern Pluralism and the Tree of Life Hypothesis." *Proceedings of the National Academy of Sciences* 104: 2043–49.

Dykuizen, D. and L. Green (1991) "Recombination in *Escherichia coli* and the Definition of Biological Species." *Journal of Bacteriology* 173: 7257–68.

Ehrlich, P. and P. Raven (1969) "Differentiation of Populations." *Science* 165: 1228–32.

Elder, C. (2008) "Biological Species Are Natural Kinds." *Southern Journal of Philosophy* 46: 339–62.

Eldredge, N. and S. J. Gould (1972) "Punctuated Equilibria: An Alternative to Phyletic Gradualism." In T. J. M. Schopf (ed.), *Models in Paleobiology*. San Francisco: Freeman Cooper, pp. 82–115.

Ereshefsky, M. (1992) "Eliminative Pluralism." *Philosophy of Science* 59: 671–90.

(2010a) "What's Wrong with the New Biological Essentialism." *Philosophy of Science* 77: 674–85.

(2010b) "Microbiology and the Species Problem." *Biology and Philosophy* 25: 553–68.

(2010c) "Darwin's Solution to the Species Problem." *Synthese* 175: 405–25.

(2011) "Mystery of Mysteries: Darwin and the Species Problem." *Cladistics* 27: 67–79.

Ferguson, J. (2002) "On the Use of Genetic Divergence for Identifying Species." *Biological Journal of the Linnean Society* 75: 509–19.

Ghiselin, M. (1969) *The Triumph of the Darwinian Method*. University of Chicago Press.

(1974) "A Radical Solution to the Species Problem." *Systematic Zoology* 23: 536–44.

Grant, V. (1981) *Plant Speciation*, 2nd edn. New York: Columbia University Press.

Griffiths, P. (2007) "The Phenomena of Homology." *Biology and Philosophy* 22: 643–58.

Hull, D. (1978) "A Matter of Individuality." *Philosophy of Science* 45: 335–60.

(1980) "Individuality and Selection." *Annual Review of Ecology and Systematics* 11: 311–32.

(1988) *Science as a Process*. University of Chicago Press.

Kitcher, P. (1984) "Species." *Philosophy of Science* 51: 308–33.

LaPorte, J. (2004) *Natural Kinds and Conceptual Change*. New York: Cambridge University Press.

Mallet, J. (1995) "A Species Definition for the Modern Synthesis." *Trends in Ecology and Evolution* 10: 294–99.

(2008) "Mayr's View of Darwin: Was Darwin Wrong about Speciation?" *Biological Journal of the Linnean Society* 95: 3–16.

Marris, E. (2007) "The Species and the Specious." *Nature* 446: 250–53.

Mayden, R. (2002) "On Biological Species, Species Concepts and Individuation in the Natural World." *Fish and Fisheries* 3: 171–96.

Mayr, E. (1969) *Principles of Systematic Zoology*. Cambridge, MA: Harvard University Press.

(1970) *Populations, Species, and Evolution*. Cambridge, MA: Harvard University Press.

(1982) *The Growth of Biological Thought. Diversity, Evolution, and Inheritance*. Cambridge, MA: Harvard University Press.

Mishler, B. (1999) "Getting Rid of Species?" In R. Wilson (ed.), *Species: New Interdisciplinary Essays*. Cambridge, MA: MIT Press, pp. 307–16.

(2003) "The Advantages of a Rank-Free Classification for Teaching and Research." *Cladistics* 19: 157.

Nesbø C., M. Dultek, and F. Doolittle (2006) "Recombination in Thermotoga: Implications for Species Concepts and Biogeography." *Genetics* 172: 759–69.

Okasha, S. (2002) "Darwinian Metaphysics: Species and the Question of Essentialism." *Synthese* 131: 191–213.

Pleijel, F. and G. Rouse (2000) "Least-Inclusive Taxonomic Unit: A New Taxonomic Concept for Biology." *Proceedings of the Royal Society B: Biological Sciences* 267: 627–30.

Podos, J. (2001) "Correlated Evolution of Morphology and Vocal Signal Structure in Darwin's Finches." *Nature* 400: 185–87.

Richards, R. (2010) *The Species Problem: A Philosophical Analysis*. New York: Cambridge University Press.

Ridley, M. (1993) *Evolution*. Cambridge, MA: Blackwell.

Rosselló-Mora, R. and R. Amann (2001) "The Species Concept for Prokaryotes." *FEMS Microbiology Reviews* 25: 39–67.

Ruse, M. (1969) Definitions of Species in Biology." *British Journal for the Philosophy of Science* 38: 225–42.

(1971) "The Species Problem: A Reply to Hull." *British Journal for the Philosophy of Science* 22: 369–71.

(1973) *The Philosophy of Biology*. London: Hutchinson.

(1987) "Biological Species: Natural Kinds, Individuals, or What?" *British Journal for the Philosophy of Science* 38: 225–42.

(1988) *Philosophy of Biology Today*. Albany NY: SUNY Press.

(1994) "Booknotes." *Biology and Philosophy* 9: 507–14.

Simpson, G. (1961) *The Principles of Animal Taxonomy*. New York: Columbia University Press.

Stackebrandt, E. (2006) "Defining Taxonomic Ranks." In M. Dworkin (ed.), *Prokaryotes: A Handbook on the Biology of Bacteria*, vol. I. New York: Springer, pp. 29–57.

Stamos, D. (2007) *Darwin and the Nature of Species*. Albany NY: SUNY Press.

Templeton, A. (1992) "The Meaning of Species and Speciation: A Genetic Perspective." In M. Ereshefsky (ed.), *The Units of Evolution*. Cambridge, MA: MIT Press, pp. 159–85.

Van Valen, L. (1976) "Ecological Species, Multispecies, and Oaks." *Taxon* 25: 233–39.

Wertz J., C. Goldstone, D. Gordon, and M. Riley (2003) "A Molecular Phylogeny of Enteric Bacteria and Implications for a Bacterial Species Concept." *Journal of Evolutionary Biology* 16: 1236–48.

Whewell, W. (1840) *The Philosophy of the Inductive Sciences, Founded upon Their History*, 2 vols. London: John W. Parker.

   (1968) *William Whewell's Theory of Scientific Method*, ed. R. Butts. Pittsburgh: University of Pittsburgh Press.

Wilson, R., M. Barker, and I. Brigandt (2009) "When Traditional Essentialism Fails: Biological Natural Kinds." *Philosophical Topics* 35: 189–215.

CHAPTER 5

DNA barcoding and taxonomic practice

David Castle

[T]he Web of Science finds over 35,000 papers (and Google Scholar over 60,000) discussing the common fruit fly, *Drosophila mela-nogaster*, described by Meigen in 1830, yet Meigen (1830) has been cited less than 60 times.

Agnarsson and Kuntner 2007

Taxonomy, the detailed study and cataloguing of the distinguishing morphological characters of the World's taxa, is an old and venerable science whose contributions to biodiversity knowledge are uncontestable. Yet support for museum- and university-based taxonomy has been withering at precisely the time when awareness of threats to global biodiversity has grown. Preserving existing, and expanding future, taxonomic knowledge is arguably intrinsically good. Certainly taxonomists would agree with this statement, but so too would people who think taxonomic knowledge is instrumentally good because it can be use to "conserve, manage, understand and enjoy the natural world" (Wheeler *et al.* 2004). Options for expanding, modernizing, and making taxonomy more relevant to practical issues are, however, rebuffed by taxonomists as undermining taxonomic science.

This chapter investigates how taxonomists have responded to DNA barcoding, a high-throughput, semi-automated method of assigning unique identifiers to taxa. The chapter begins with an overview of barcoding, its aims and methods, before turning to a discussion of the three main objections brought forward by taxonomists. These objections concern the methods and technology, the use of barcoding in biodiversity science, and the expectations of how barcoding will expand contributors and users of taxonomy. In this chapter it is argued that barcoding is an evolving method, and assessments of barcoding's weaknesses, and improvements upon its strengths, are ongoing. The reasons taxonomists cite for rebuffing barcoding, however, often have more to do with protecting past taxonomic practices than they do with barcoding itself; it is just

that barcoding exemplifies to taxonomists many perceived threats they
most fear.

DNA barcoding

DNA barcoding is a method by which groups of organisms can be differentiated by comparing short, standardized regions of DNA. Barcodes
are created from a tissue sample of a field or museum specimen, extract-
ing the DNA, amplifying the barcode region, and then sequencing it.
The sequence information is combined with details about the specimen
to produce a unique record included in an already confirmed species
identification, or part of a library that can support species identifications
(Ratnasingham and Hebert 2007). Barcoding is a high-throughput, partially automated biodiversity informatics platform that that functions
internationally through networked country or regional nodes, with a core
sequencing and database facility at the Biodiversity Institute of Ontario.

Barcoding as a method of biological identification is described by Paul
Hebert and colleagues (Hebert *et al.* 2003) as a way of identifying the
10–15 million species thought to exist. In addition to being time-consum-
ing, morphological identification struggles with phenotypic and genetic
variation, cryptic species, life-stage differences, and the need for expertise
to ensure accurate identifications. Furthermore, "microgenomic identification systems" can address the "limitations inherent in morphology-based
identification systems and the dwindling pool of taxonomists." As they
point out, the idea of using DNA to identify organisms is not a new concept. The novel contribution, they argue, is to take "the discrimination
of life's diversity from a combinatorial perspective." Whereas Universal
Product Codes permit 100 billion unique barcode identifiers commonly
seen on commercial products, 15 nucleotide positions generate 4^{15} – or 1
billion – codes. Given easily obtainable sequence information, as more
positions are analyzed, resolution goes up and error rates go down. They
suggest that, "[T]hese sequences can be viewed as genetic 'barcodes' that
are embedded in every cell."

The choice of the standard region to sequence has consequences not
only for the technical feasibility of isolating and amplifying the DNA, but
also underpins the use of barcodes to resolve taxa. While it would obviously be preferable to have a standard region for barcoding that would
work for all organisms, no universal region has been identified, nor is it
likely to be given evolutionary divergence. For animals, the cytochrome
c oxidase (CO1) is used because the universal primers for the gene are

"robust" and because it "appears to possess a greater range of phylogenetic signal than any other mitochondrial gene" (Hebert *et al.* 2003). From a practical standpoint, the rate of COI evolution and associated amino acid changes enables identification of species-level and higher taxa, as well as the disambiguation of closely allied species and sub-species-level phylo-geographic groups.

The low level of nucleotide substitution in plants has meant that mito-chondrial regions such as COI would not be useful for barcoding plants (Fazekas *et al.* 2008). Instead, chloroplast DNA regions have been consid-ered for a number of years, but the identification of the regions has been a challenge. Similar to the considerations that drove the selection of COI, plant regions have been selected on the basis of how well the DNA region can be amplified and sequenced, patterns of genetic variation and sequence quality, and success corroborating species identifications with barcodes – that is, discriminatory power (Hollingsworth 2008). A combination of two sequences, *matK* and *rbcL*, has been accepted as the core barcodes for plants (CBOL Plant Working Group 2009). Although there are unresolved limi-tations with the pair of accepted regions, the focus is now on building up a large barcode database to understand better how plant barcoding works in practice across a wide range of taxa before choosing new regions to add or substitute (Hollingsworth *et al.* 2011). There are instances in which hav-ing an additional region has proved beneficial for differentiating specimens from species-rich communities (Kress *et al.* 2009).

Fungal diversity is particularly challenging given the worldwide abun-dance of fungi and their complex biology. As has been observed, "[B]ecause most fungi are microscopic and inconspicuous and many are unculturable, robust, universal primers must be available to detect a truly representative profile" (Schoch *et al.* 2012). Feasibility studies were under-taken to assess whether the COI could be the standard barcode region for fungi, and initially on the basis of analyzing *Penicillium*, COI appeared to be adequately diverse to differentiate taxa (Seifert *et al.* 2007). Other stud-ies demonstrated while COI is potentially sufficiently diverse in fungi to be a standard barcode region, fungi other than *Penicillium* have too many, and often large, introns in the region (Vialle *et al.* 2009). The nuclear ribosomal internal transcribed space (ITS) region has been used in genetic identification of fungi for decades, and its suitability as a barcoding marker has been favorably assessed (Schoch *et al.* 2012). The ITS region is on a par for accuracy with the two-region approach for plants (about 70 percent), and its ubiquity in fungi and relative ease of amplifying and sequencing has led to its becoming the standard.

Because fungi, like bacteria, are often difficult to differentiate morphologically, taxonomists have long used sequence divergence to differentiate lineages. Operational taxonomic units (OTUs) arise from
applications of the sequence divergence benchmark, a process that
creates operational terminal nodes in genetic phylogenies. Automated
sequencing of barcodes, particularly with increasing throughput of specimens, presents challenges for creating algorithms to identify OTUs.
Methods for animal barcode OTUs (Blaxter *et al.* 2005) have been developed, and several other algorithms have been developed since that vary
in the way that they handle sequence differences (number of nucleotides
considered and distances between them) and the underlying statistical
approaches. Five algorithms have recently been compared on eight datasets to assess how well they perform in making a taxonomic assignments
(Ratnasingham and Hebert 2013). Seven of the eight datasets were from
well-described North American lepidoptera – "the taxonomic assign-
ment for each record in these trial datasets was treated as a 'truth'" –
but with the proviso that even well-studied groups might have imperfect
taxonomic assignments.

A system of unique sequence record identifiers called the Barcode Index
Number (BIN) creates a registry of the specimen sequence data (> 500
bp) combined with the raw sequence trace file, information about the
specimen collection and geospatial data. The BIN system can be queried
and annotated by users through an online portal to the Barcode of Life
Data System (BOLD). BOLD is a workbench consisting of BIN records.
Each database record has seven elements: (1) a species name, even if pro-
visional; (2) voucher data; (3) collection record; (4) the specimen identi-
fier; (5) sequence information; (6) primer information; and (7) trace files
(Ratnasingham and Hebert 2007). BOLD records can be integrated into
GenBank. At the time of writing, BOLD contains 2,012,391 sequenced
specimens with a BIN of which there are 172,280 named species.

The barcoding initiative is focused on understanding the diversity of
life (Hanner and Gregory 2007). The technology is promoted as being
capable of generating a library of barcodes that will create clarity about
how diverse life is, while providing data for identifying organisms and
differentiating them genetically from morphologically indistinct relatives.
In this respect, DNA barcoding is an adjunct to traditional taxonomy,
particularly molecular taxonomic methods. Recently, it has been argued
that DNA barcoding is also the underlying method of all 'DNA metasystematics' in which genetic information is used to identify individuals from
samples in which there are mixtures of DNA, for example in complex soil,

water, or insect-trap collections (Hajibabaei 2012) of the sort one might find in biodiversity-rich tropical environments (Kress *et al.* 2009). Within this framework,

> genetic information is obtained directly from environmental samples to understand biodiversity at different levels of organization and taxonomic groups. Metasystematics studies can provide species-level biodiversity information through the use of standardized DNA barcode markers that are linked to reference sequence libraries, voucher specimens and, ultimately, all available taxonomic knowledge gained through past investigations. Different DNA markers can provide deeper phylogenetic diversity or functional diversity measures (i.e., genes involved in various path-ways). (Hajibabaei 2012)

Barcoding, as the basis of DNA metasystematics, could consequently promote interdisciplinary research into community ecology and molecular genetic evolution for example.

The potential for "profound socioeconomic utility" (Hajibabaei 2012) of biodiversity monitoring and the use of barcoding to study whole communities has stimulated many applications of DNA barcoding. Barcoding can be used to identify medicinal plants (Chen *et al.* 2010), which can be useful for new drug discovery as well as monitoring plants and extracts that are traded internationally. Biodiversity studies uphold the Convention on Biological Diversity (Vernooy *et al.* 2010) and barcoding can be used to identify resources or resolve disputes related to access and benefits-sharing schemes. Barcoding has also been incorporated as a tool for identifying invasive species (Armstrong 2010). Illicit trade practices, for example the selling of illegal bushmeat, can be monitored using barcodes (Eaton *et al.* 2010), as can the marketplace substitution of low-value fish (Wong and Hanner 2008). Barcoding has also been used stimulate interest in biodiversity science within schools (Schoch *et al.* 2012), and recently Google made a large investment in supporting the use of barcodes to monitor species at risk (CITES 2012).

Critiques of DNA barcoding

Barcoding attracts considerable and sustained criticism from a wide variety of perspectives and sources. Most of this criticism revolves around three related issues. The first concerns how barcoding is done – the underlying technology and associated methods. The second line of criticism is partly based on the first, arguing that barcoding methods and practices have been positioned in ways hostile to traditional taxonomy. The third line of criticism argues that barcoding's ambition to change taxonomic practice by mobilizing the public has major shortcomings.

Underlying technology and associated methods

The selection of standard regions for DNA barcoding has generated discussion (Prendini 2005) ever since the CO1 region for barcoding animals was described in Hebert *et al.* (2003). An early detected problem with using the CO1 region with amphibians requires other sequences to be used (Vences *et al.* 2005), undermining hopes that a ubiquitous region would be used for animals. There has been a rise in the number of additional sequences used in animal barcoding in cases where there are insufficient differences in the barcode region between closely related species (Taylor and Harris 2012). As described above, a candidate region for all of life was never identified, a fact which has led to the selection of two regions in plants and a ribosomal region in fungi. These regions are generally thought to be underperforming by about 25 percent, meaning that when well-described species are evaluated, just better than 70 percent of them can be distinguished. A higher resolving power would be preferable, of course.

The strongest criticism about the technology and method itself raises doubts about the use of distances between varying nucleotide sequences to differentiate specimens. Regarding the use of distances, it has been argued that they are an arbitrary criterion that is difficult to integrate with character-based taxonomy, and distances are not guaranteed to give nearest neighbors or to successfully delineate taxa (DeSalle *et al.* 2005). DeSalle *et al.* also argue that distances, which are of course measurable, are not the sort of data that should be used to construct phylogenies. In particular, they argue against the "use of single gene trees as evidence of phylogenetic relationships," preferring instead an approach that, if it uses barcodes at all, would combine them with character-based methods (DeSalle *et al.* 2005).

A more recent criticism of barcoding confronts the fact that barcoding was developed, and continues to use, Sanger sequencing when next-generation sequencing could not only be used to read several DNA regions, but could in the not too distant future be able to do whole-genome sequencing quickly and inexpensively (Taylor and Harris 2012). The small size of the barcoding regions currently are responsible in part for driving the scale of the BOLD database system; larger regions would have required greater data storage and higher computational power. To this, Taylor and Harris object that the BOLD system must be able to handle next-generation sequencing:

If nothing else, this will provide an efficiently organised storage area for longer/full sequence data destined for species identification until it can be verified, encouraging further deposition of tissue specimens. Failure to bring BOLD up to speed with NGS would represent another seeming counterproductive and stubborn stance in the face of scientific advancement. (Taylor and Harris 2012)

Studies using next-generation sequencing are becoming numerous as new sequencing platforms become more available, for example Hibert *et al.* (2013); Wilkening *et al.* (2013).

Barcoding and traditional taxonomy

DNA barcoding as developed and practiced by the Consortium for the Barcode of Life (CBOL) and the International Barcoding of Life project (iBOL) has existed for a decade, and came into existence during a period in which university- and museum-based taxonomy has experienced a funding decline while insights into the scale of the biodiversity crises have grown (Wilson 1985). As discussed previously, molecular taxonomic methods have existed for decades, but barcoding is generally not regarded as having direct continuity with these methods, partly because it is a semi-automated high-throughput process, but also because it is viewed as an informatics-driven approach to biodiversity. As a result, the relationship between barcoding and traditional, morphology-based taxonomy has been a source of dispute among barcoding proponents and detractors. There are two central issues; the first is whether barcoding can be seen to be a complementary activity to morphological taxonomy, and the second is whether the spread of barcoding initiatives worldwide undermines traditional taxonomy.

Hebert *et al.* (2003) said that barcoding could be used for species identification, not just in cases where morphological approaches would be difficult or impossible, but more generally: "The general ease of species diagnosis reveals one of the great values of a DNA-based approach to identification. Newly encountered species will ordinarily signal their presence by their genetic divergence from known members of the assemblage." Hebert's perspective about the relationship between barcoding and taxonomy was subsequently comprehensive:

In addition to assigning specimens to known species, DNA barcoding will accelerate the pace of species discovery by allowing taxonomists to rapidly sort specimens and by highlighting divergent taxa that may represent new species. By augmenting their capabilities in these ways, DNA barcoding offers taxonomists the opportunity to greatly expand, and eventually complete, a global inventory of life's diversity. (Hebert and Gregory 2005)

The choice of 'inventory' is notable, and elsewhere in Hebert and Gregory (2005) it is remarked that although "DNA barcoding will not create the 'encyclopedia of life,' it will generate its index and a table of contents." History, they conclude, "may view the DNA barcoding enterprise as one that not only enhanced access to taxonomic information, but also strengthened alliances among all those with interests in the documentation, understanding, and preservation of biodiversity – an exciting prospect indeed."

This prospect has failed to excite many detractors of barcoding, who have been persistent in expressing two main concerns over the past decade. The first concern is that barcoding diverts resources away from traditional, morphology-based taxonomy. When barcoding is portrayed as having "the potential to accelerate our discovery of new species, improve the quality of taxonomic information, and make this information readily available to nontaxonomists and researchers outside of major collection centers" (Miller 2007), the conclusion drawn by some taxonomists is that barcoding will not bolster their efforts and raise their profile among science funders and the public, but will instead usurp their resource base and further erode traditional taxonomic practice. Barcoding could rearrange funding priorities by receiving large-scale funding as big science "and as a result is viewed (wrongly) as a modernized taxonomy, [and] will in fact begin supplanting taxonomic projects" and "it is naïve to believe that the drive toward universal barcoding will not bring with it a radical shift in overall funding priorities" (Ebach and Holdrege 2005). The impact will be noticeable in institutions like museums struggling to curate collections and maintain skilled staff:

> In a funding climate focused on promoting sexy new high-output "solutions" to global problems, a scientific field that progresses by investing much time, energy, and funding into training taxonomists, doing careful fieldwork, and carrying out detailed morphological studies may seem outmoded. According to this view, taxonomists soon will become fossils in the strata of scientific evolution themselves. (Ebach and Holdrege 2005)

The taxonomic impediment, long discussed before the advent of DNA barcoding, "is a problem compounded of several deficiencies: an acute shortage of taxonomic experts, uneven distribution of resources (reference collections, literature) in the world, inefficient practices in conducting taxonomic research, and taxonomic products with limited utility to likely consumers" (Rodman and Cody 2003). Barcoding is criticized for not helping to overcome the impediment and, worse, taking resources away from taxonomists that would help to overcome it.

The second concern is that if barcoding is considered a part of taxonomic practice, it will ultimately undermine the quality of taxonomic work. At issue is whether the use of short, standardized regions of highly conserved DNA can be used to differentiate species. Miller (2007) suggests the problem is that barcoding reignites very old debates about what species are, and how they are identified using different kinds of data and methods. As he says, "DNA barcoding, like previous applications of new categories of characters, creates the challenge of integrating new data into an established knowledge framework" (Miller 2007). Where there is strong convergence between accepted morphological descriptions of species and the separation of specimens into species using barcode data (Hajibabaei *et al.* 2006), barcoding appears to be validated as a rapid method for distinguishing taxa. Yet because single characters are generally seen as inadequate for making species identifications, there has generally been a need to find characters – morphological or genetic – that indicate interspecific divergence. The debate "continues over the philosophical basis for species concepts and how to operationalize these concepts into a system for recognizing species" (Miller 2007).

This debate generally falls along two distinct lines. Many advocates of barcoding think that barcoding can support traditional taxonomy – Hajibabaei *et al.*'s (2006) barcoding analysis led to the disambiguation of 13 species, subsequently confirmed by morphological analysis. Without having to defend a strong position about a species definition, barcoding can be seen as one among many methods for distinguishing taxa, but not the sole arbiter of taxonomic divisions. Others raise doubts whether barcoding contributes anything of use to taxonomy. Ebach and Holdrege (2005), for example, quote Sydney Brenner, who wrote of the Human Genome Project that we now "know that the genome sequence is only the beginning and that a deduced amino acid sequence is not a target for anything unless we know how it participates in the physiological processes in our bodies" (Brenner 2003). That is, barcoding generates 'information' and 'not knowledge', and runs the risk of not identifying 'real species', an outcome that would not only threaten barcoding but poses a greater threat in Ebach and Holdrege's view to taxonomy itself. The information–knowledge distinction is often difficult to draw in actual scientific practice, but advocates of barcoding do not think that uninterpreted barcodes constitute knowledge (Schindel and Miller 2005).

A perspective of taxonomy in which skilled experts make morphology-based identifications of real species reasserts the role and expertise of the university- or museum-based expert. Consequently, suggesting that a "DNA-based future can herald several possible fates for the Linnaean system – from outright dismissal to revitalization" (Hebert and Barrett 2005) is bound to excite controversy. As will be discussed below, authors like Taylor and Harris (2012) cite the quotation out of context and give the impression that barcoding is an 'anything goes' science that embraces the opportunity to overthrow non-molecular taxonomy. While criticism of Linnaean hierarchies, and the speed with which species are identified using the system, are well established (Ereshefsky 2000), the concern is that 'DNA species' will replace morphological species. In addition to the objection about information and knowledge, the consequences of using a semi-automated process on taxonomy are considered problematic. It raises for barcoding the equivalent of another of Brenner's suggestions about the Human Genome Project – "that project leaders parcel out the job to prisoners as punishment – the more heinous the crime, the bigger the chromosome they would have to decipher" (Roberts 2001).

One suggestion to address the taxonomic impediment is to provide a greater number of species names from automated processes, thereby giving taxonomists increased numbers of species to describe – a process which precisely inverts how morphological taxonomy normally applies species names (Evenhuis 2007). One response reasserts the role of assigning species names *following* inferences about characters and taxic evolution, reasserting that "even though 'quantity' is presently needed from taxonomy, it cannot come at the expense of 'quality', as taxonomic names corresponding to flawed hypotheses of biological entities will compromise the reliability of systematic information for society" (de Carvalho *et al.* 2008). What would contribute to quality systematics, therefore, is taxonomy that strives for genuine monophyly, with phylogenetic information included as possible.

Without this focus for taxonomy, de Carvalho *et al.* object that taxonomy loses its epistemological underpinnings and realism, and likens automated approaches to generating names to pheneticism:

> With the availability of calculators and computers in the late 1950s and early 1960s, there was great optimism concerning the development of an operational, quicker, and "non-subjective" taxonomy, strongly propagated by non-evolutionary taxonomists. Twenty-five years later, the founders of this approach had to abandon its principles because they realized that the taxa they proposed, both at the species level and above it, had no real connection with nature. (de Carvalho *et al.* 2008)

Barcoding has been likened to phenetics more than once: "proponents of DNA barcoding argue that these problems can all be solved quickly and efficiently by replacing morphological methods ... with molecular techniques" (Hamilton and Wheeler 2008). On this view, using barcodes derived from short regions of DNA, combined with measures of nucleotide distance and the support of statistical algorithms, is viewed as a recapitulation of a kind of numerical taxonomy (Sokal and Sneath 1963). Proponents of barcoding disagree that molecular markers are arbitrary characters that have no phylogenetic significance or meaning in systematics.

The entire 'barcoding enterprise' of individuals, institutions, organizations, technology vendors, and end users has been labeled a 'pseudoscience', one that is 'anti-intellectual' because it closes down rather than opens up scientific enquiry and insight (Ebach and de Carvalho 2010). The crux of the argument is that barcoding does not live up to the inviolable scientific norms of morphology-based taxonomy:

> A database, no matter how well atomized or sorted, cannot replace the knowledge of a taxonomist. As a tool the database is invaluable, but as a source of knowledge of a particular group, it is inadequate. When databases, for instance, start to out-number taxonomists the knowledge is not replaced – only nomenclature, descriptions and hypothetical phylogenetic relationships are recorded. If we wish to increase the productivity in taxonomy then quite obviously we increase the number of taxonomists and demand from them products of high quality. This adheres to best practice. (Ebach and de Carvalho 2010)

Like the criticisms of barcoding as a new variant of phenetics or numerical taxonomy, this objection searches for evidence that the underlying motivation of barcoding is to supplant, not augment, traditional taxonomic methods. Otherwise, it would be difficult to conclude that "DNA barcoding is at best pseudoscientific and held together by a fundamental belief that molecules speak the 'truth' and technology 'saves'" (Ebach and de Carvalho 2010, 174).

Mobilizing the public, or 'publics', of biodiversity science and barcoding

The biodiversity crises conjoined with the taxonomic impediment have prompted biodiversity scientists to seek new means of collecting and identifying the world's biodiversity. Wilson estimates there are approximately 6,000 taxonomists working worldwide, having identified just over 1.5 million species with as many as 10 million species thought to exist (Wilson 2004). Localized urbanization and resource development in biodiverse regions, and diffuse but no less significant effects of climate change, are leading to massive extinctions. A sense of urgency has developed within biodiversity science that these trends must be reversed – hence the biodiversity crises – but given that a small fraction of the estimated number of species has been identified, there is a new mission to collected specimens as rapidly as possible. Otherwise, ignorance will prevail about how much biodiversity is at risk or how much has been lost. One approach to speed up collections in biodiversity hotspots has been the development of 'parataxonomists', individuals who are given basic training in the collection and preservation of specimens, and who can thereby support rapid assessments of biodiversity (Oliver and Beattie 1993).

Janzen, who worked in Costa Rica to establish the National Biodiversity Institute (INBio), has been an advocate of parataxonomy for decades and well in advance of the advent of COı barcoding techniques in 2003 (Janzen 1991). Janzen's primary objective was to collect specimens, and he argued that the 100–200 parataxonomists trained through INBio would bolster Costa Rican collections (Tangley 1990). The effort worked: "Just six months after they completed the course, members of the first class had collected four times as many insect specimens than had been put in the national collections during the past 100 years." Janzen's interest in rapid collection development has made him an early adopter of barcoding (Janzen 2004) and now a staunch advocate of the utility of barcoding for field collections and first-pass identifications. Barcoding is seen as a key element in overcoming humanity's bioilliteracy: "True bioliteracy is being able to link what humanity knows to the biodiversity in hand, eye, or mouth, and build on it" (Janzen 2010). Janzen advocates for a global movement of specimen collection and barcoding to raise awareness and respond to the biodiversity crisis, but also to overcome the taxonomic impediment by using barcoding to help screen, perhaps one day using handheld devices, specimens in the field. If they are already barcoded, this can be verified by remote connection to a barcode library and the parataxonomist can move on; if not, they collect the specimen.

> Who will fill that barcode library? The biodiversity priests and acolytes will. There are many of them, easily 100,000-strong across the globe. There are an amazing number of people who "know" and handle a subset of the world's biodiversity: taxonomists, parataxonomists, hobbyists, government agencies, teachers, owners, nature lovers, biodiversity prospectors, game wardens, environment monitors, conservationists, etc. They will be pleased to submit an "identified" bit of each species that they know or encounter,

so as to have its DNA barcode in the barcode public library, so long as they do not have to pay the cost of putting it there. (Janzen 2010)

This inclusive, worldwide movement of barcoding-led specimen collection and identification is appealing to many biodiversity scientists, and has piqued Google's interest in supporting the project on endangered species (CITES 2012). Social scientists, and taxonomists, have not, however, embraced this vision with the same fervor as has Janzen.

The character of the individuals and the public that embraces barcoding is not very clear, particularly with respect to how their encounters with biodiversity science and barcoding technologies will unfold. One group of authors has attempted to understand what kinds of public engagements with science were envisioned, and how those encounters would transpire by understanding how the barcoding community is "discursively articulating an abstract generalized and global 'public'" and how this "rather nebulous 'general public'" can be distinguished from "two further 'publics' imagined for this techno-science" (Ellis *et al.* 2010). The groups they identify are, first, the taxonomic community "for whom industrial-style rapid species diagnosis is freeing up valuable time for more complex pursuits than simple organism identification"; second, the "non-taxonomist professionals with a need for rapid species diagnosis"; and third, the global or "general public." The authors contrast portrayals of bioilliterate citizens with the barcoding community's visions of the potential for a global, bioliterate public supported by barcoding and biodiversity science. Yet as decades of research on the complexities and uncertainties in the public understanding of science suggest, no such transition in attitudes, beliefs, and interaction with science, technology, and the natural world are likely to come easily or quickly.

The aspiration of fulfilling the needs of an inquisitive, bio-enlightened public maintains a long tradition of science education and public museums and zoos. Barcoding is promoted as allowing "a day to be envisioned when every curious mind, from professional biologist to schoolchildren, will have easy access to the names and biological attributes of any species on the planet" (Hebert and Gregory 2005). More crucial to the direct operations of barcoding, however, is the development and deployment of parataxonomists. The prospect of parataxonomy holds little appeal for taxonomists, who think that they will be left with "unidentifiable barcodes" or "revising problematic taxonomic groups in order to streamline classification" (Ebach 2011). If 'parataxonomy' has a somewhat dignified ring to it, 'do-it-yourself biology' (Godfray 2007), analogous to DIY electronics and 'garage biology', was certain to draw fire. Godfray's invocation of Linnaeus for an 'information age' was interpreted as a "revisionist strategy employed to recruit that great luminary in support of Godfray's own cause", namely a "concerted directive to discredit 'traditional' or 'established' systematics" (de Carvalho *et al.* 2008).

Parataxonomists, as one group or 'public', are connected to the third 'public' described by Ellis *et al.* – the end users of barcoding information. Critics of barcoding do not endorse what they see as a vertically integrated supply chain of specimens – samples – metadata – barcodes – informatics – identifications – applications. Taxonomy is contrasted with "identification service" when it ought to be an "independent science" (Hamilton and Wheeler 2008). This point of view is amplified by Ebach and de Carvalho, in a response to Miller (2007):

> Taxonomy is not an identification service and does not answer to "economic development" or "local, national and international user communities." In order to expedite taxonomy, greater investment into training taxonomists is needed as well as a greater number of available positions. A long-term vision, such as identifying all the species on Earth, requires long-term investments. Unfortunately, DNA Barcoding is sensationalism that promotes new technologies, which are not able to perform the same functions as taxonomists, is damaging to systematics and taxonomy. (Ebach and de Carvalho 2010)

Interestingly, while social scientists have striven to disambiguate the 'publics' associated with DNA barcoding to be able to understand the dynamics between them, taxonomists responding to the potential for uptake of barcoding raise arguments against the vertical integration of barcoding that collapse parataxonomists, end users, and the general public together.

Barcoding and taxonomy: transformation or stalemate?

Spirited challenges to barcoding have been launched in the decade since the Hebert paper recommended the CO1 region for animal barcoding in 2003. These challenges focus on whether barcoding is scientifically and technologically grounded, whether barcoding is a Trojan horse for traditional, morphology-based taxonomy, and whether a barcoding movement involving the public in different ways is a disservice to biodiversity science generally and to taxonomy in particular. The question is whether these challenges are achieving the protection and promotion of taxonomy that is desired, or whether they are further isolating traditional taxonomy.

Through the course of the last six decades it is fair to say that biology has become increasingly focused on the molecular level, has widely embraced various chemical and genetic and other chemical high-throughput techniques, and has become increasingly driven by informatics. All of this is occurring as various branches of organismic and ecosystem biology have not experienced the same ascendancy in the broader scientific agenda, and in some cases have experienced significant contraction in institutional and funding support. While taxonomy has stalled, resources for phylogenetic reconstructions have surged. Yet as meritorious as taxonomic science may be, it is something of a lump of labor fallacy to suggest that support for phylogenetics is paid for by reduced taxonomic effort. In any event, the conjecture that barcoding is financed by reductions in taxonomic science could only be proved by documenting that Peter is being robbed to pay Paul, or by testing a counterfactual situation in which molecular techniques never gain purchase in biological science.

Not all scientists agree that taxonomy should be left to its own devices, decoupled from local issues that were at issue between Miller and Ebach above, and paying little or no attention to practical issues of species identification for a wide range of purposes including single-species identification or, more broadly, biodiversity monitoring of ecosystem services (Kim and Byrne 2006). Kim and Byrne challenge taxonomists, whom they see as being unwilling to engage with interdisciplinary biodiversity science, and ecologists, whom they see as having under-appreciated the need for good taxonomic data, to unite and produce a renewed biodiversity science. This science would focus on regional, all-taxa assessments, would become networked and capable of providing local taxonomic services, and would advance a training mandate for a "new generation of broad-thinking, transdisciplinary scientists" (Kim and Byrne 2006). There are others who have realized that the future of taxonomy is likely to be dependent on working across disciplines (Smith *et al.* 2006), and advice has been given to future taxonomists that the discipline and individuals' futures depend on finding new ways to ensure that taxonomic work is published (Agnarsson and Kuntner 2007).

Even critics of barcoding have argued that "[r]ather than replacing tradition taxonomy, DNA barcoding has actually reinforced the need for qualified taxonomists by producing sequence data that needs to be paired with a verified morphological type specimen" (Taylor and Harris 2012). Taylor and Harris rightly follow up this remark by pointing out that next-generation sequencing might create a data backlog, but they would have to acknowledge that is just the point – it would expose the shortage of taxonomists for what it is. In making this remark they cite, but do not examine, however, a paper that explored the deficiencies of morphological taxonomy in comparison with barcoding techniques that have proved in many instances to excel in species identification (Packer *et al.* 2009). As a recent report observes:

> Contrary to the view of some ... the use of DNA sequences as characters does not threaten to make taxonomists redundant. Rather, it is empowering them to focus more time on the interesting research questions, while delivering identification support and biodiversity assessments to more users more quickly in many more places. (House of Lords Science and Technology Committee 2008)

A consensus is indeed emerging that saying that barcoding can replace taxonomists is an overstatement, as is any statement that barcoding is pseudo-science; both claims are themselves obstacles to improving biodiversity science.

In their discussion of how taxonomy could move from 'impediment to expedient', Wheeler *et al.* (2004) argue that taxonomy needs to incorporate a "cyber-infrastructure," real-time remote microscopy, taxonomists in the classroom, worldwide access to taxonomic literature, and teams of "taxonomists to speed species exploration" (Wheeler *et al.* 2004). This is not incongruent with a description of the role of barcoding in taxonomy and biodiversity science: "[W]e are not defending 'traditional taxonomy' here, but instead we argue that the real cutting-edge future for systematics and biodiversity research is *integrative taxonomy* which uses a large number or characters including DNA and many other types of data, to delimit, discover, and identify meaningful, natural species and taxa at all levels" (Will *et al.* 2005). Integrative taxonomy aims to combine the frequently separated tasks of delimiting and classifying species, and in this respect will unite taxonomists with non-taxonomists (Dayrat 2005).

Conclusion

Claims to the effect that "the ground has shifted, and it is now DNA barcoding, rather than traditional taxonomy, which runs the risk of becoming irrelevant if it refuses to embrace change" (Taylor and Harris 2012) make good rhetoric but do not track the expansion of barcoding and refinement of the method over the last decade. Barcoding is here to stay and it will evolve over time; the choice for taxonomists is to ignore it or become involved (Stevens *et al.* 2011). On balance it would be perilous for taxonomy, and deleterious for biodiversity sciences for taxonomists to be uninvolved. There is an equal or greater risk, particularly concerning the question of resources for taxonomists, to appear to be disengaged with integrative taxonomy involving barcoding. Indeed, many taxonomists based in universities, in museums, and across industrialized and developing countries are participants in biodiversity science that involves barcoding. In the long run, it is better for science and for biodiversity if scientists realize that within their changing community of practice they have three basic options: they can give voice to their concerns, they can exit the debate altogether, or they can stay parochially loyal to past practice in the face of inevitable change (Hirschmann 1970).

REFERENCES

Agnarsson, Ingi and Matjae Kuntner (2007) "Taxonomy in a Changing World: Seeking Solutions for a Science in Crisis." *Systematic Biology* 56 (3): 531–39.

Armstrong, K. (2010) "DNA Barcoding: A New Module in New Zealand's Plant Biosecurity Diagnostic Toolbox." *EPPO Bulletin* 40 (1): 91–100.

Blaxter, Mark, J. Mann, T. Chapman, F. Thomas, C. Whitton, R. Floyd *et al.* (2005) "Defining Operational Taxonomic Units Using DNA Barcode Data." *Philosophical Transactions of the Royal Society B: Biological Sciences* 360 (1462): 1935–43.

Brenner, Sydney (2003) "Humanity as the Model System." *Science* 302 (5645): 533.

CBOL Plant Working Group (2009) "A DNA Barcode for Land Plants." *Proceedings of the National Academy of Sciences* 106 (31): 12794–97.

Chen, Shilin, Hui Yao, Jianping Han, Chang Liu, Jingyuan Song, Linchun Shi *et al.* (2010) "Validation of the ITS2 Region as a Novel DNA Barcode for Identifying Medicinal Plant Species." *PLoS ONE* 5 (1): e8613.

CITES (2012) "Cbol Receives $3 Million Grant from Google Foundation to Barcode Endangered Species." www.cites.org/eng/news/sundry/2012/20121206_CBOL.php (accessed May 10, 2013).

Dayrat, Benoit (2005) "Towards Integrative Taxonomy." *Biological Journal of the Linnean Society* 85 (3): 407–15.

De Carvalho, Marcelo R., Flávio A. Bockmann, Dalton S. Amorim, and Carlos Roberto F. Brandão (2008) "Systematics Must Embrace Comparative Biology and Evolution, Not Speed and Automation." *Evolutionary Biology* 35 (2): 150–57.

DeSalle, Rob, Mary G. Egan, and Mark Siddall (2005) "The Unholy Trinity: Taxonomy, Species Delimitation and DNA Barcoding." *Philosophical Transactions of the Royal Society B: Biological Sciences* 360 (1462): 1905–16.

Eaton, Mitchell J., Greta L. Meyers, Sergios-Orestis Kolokotronis, Matthew S. Leslie, Andrew P. Martin, and George Amato (2010) "Barcoding Bushmeat: Molecular Identification of Central African and South American Harvested Vertebrates." *Conservation Genetics* 11 (4): 1389–404.

Ebach, Malte C. (2011) "Taxonomy and the DNA Barcoding Enterprise." *Zootaxa* 2742: 67–68.

Ebach, Malte C. and Craig Holdrege (2005) "More Taxonomy, Not DNA Barcoding." *BioScience* 55 (10): 823–24.

Ebach, Malte C. and Marcelo R. de Carvalho (2010) "Anti-Intellectualism in the DNA Barcoding Enterprise." *Zoologia* 27 (2): 165–78.

Ellis, Rebecca, Claire Waterton, and Brian Wynne (2010) "Taxonomy, Biodiversity and Their Publics in Twenty-First-Century DNA Barcoding." *Public Understanding of Science* 19 (4): 497–512.

Ereshefsky, Marc (2000) *The Poverty of the Linnaean Hierarchy: A Philosophical Study of Biological Taxonomy*. Cambridge University Press.

Evenhuis, Neal L. (2007) "Helping Solve the Taxonomic Impediment: Completing the Eight Steps to Total Enlightenment and Taxonomic Nirvana." *Zootaxa* 1407: 3–12.

Fazekas, Aron J., Kevin S. Burgess, Prasad R. Kesanakurti, Sean W. Graham, Steven G. Newmaster, Brian C. Husband *et al.* (2008) "Multiple Multilocus DNA Barcodes from the Plastid Genome Discriminate Plant Species Equally Well." *PLoS ONE* 3 (7): e2802.

Godfray, H. C. J. (2007) "Linnaeus in the Information Age." *Nature* 446 (7133): 259–60.

Hajibabaei, Mehrdad (2012) "The Golden Age of DNA Metasystematics." *Trends in Genetics* 28 (11): 535–37.

Hajibabaei, Mehrdad, Daniel H. Janzen, John M. Burns, Winnie Hallwachs, and Paul D. N. Hebert (2006) "DNA Barcodes Distinguish Species of Tropical Lepidoptera." *Proceedings of the National Academy of Sciences* 103 (4): 968–71.

Hamilton, Andrew and Quentin D. Wheeler (2008) "Taxonomy and Why History of Science Matters for Science." *Isis* 99 (2): 331–40.

Hanner, R. H. and T. R. Gregory (2007) "Genomic Diversity Research and the Role of Biorepositories." *Cell Preservation Technology* 5: 93–103.

Hebert, Paul D. N. and Rowan D. H. Barrett (2005) "Reply to the Comment by L. Prendini on 'Identifying Spiders through DNA Barcodes'." *Canadian Journal of Zoology* 83 (3): 505–6.

Hebert, P. D. and T. R. Gregory (2005) "The Promise of DNA Barcoding for Taxonomy." *Systematic Biology* 54 (5): 852–59.

Hebert, P. D., S. Ratnasingham, and J. R. deWaard (2003) "Barcoding Animal Life: Cytochrome *c* Oxidase Subunit 1 Divergences among Closely Related Species." *Proceedings of the Royal Society B: Biological Sciences* 270 (Suppl. 1): S96–S99.

Hibert, Fabrice, Pierre Taberlet, Jérôme Chave, Caroline Scotti-Saintagne, Daniel Sabatier, and Cécile Richard-Hansen (2013) "Unveiling the Diet of Elusive Rainforest Herbivores in Next Generation Sequencing Era? The Tapir as a Case Study." *PLoS ONE* 8 (4): e60799.

Hirschman, Albert O. (1970) *Exit, Voice, and Loyalty: Responses to Decline in Firms, Organizations, and States*. Cambridge, MA: Harvard University Press.

Hollingsworth, Peter M. (2008) "DNA Barcoding Plants in Biodiversity Hot Spots: Progress and Outstanding Questions." *Heredity* 101 (1): 1–2.

Hollingsworth, Peter M., Sean W. Graham, and Damon P. Little (2011) "Choosing and Using a Plant DNA Barcode." *PLoS ONE* 6 (5): e19254.

House of Lords Science and Technology Committee (2008) *Systematics and Taxonomy: Follow-up Report with Evidence.* London: The Stationery Office.

Janzen, Daniel H. (1991) "How to Save Tropical Biodiversity: The National Biodiversity Institute of Costa Rica." *American Entomologist* 37 (3): 159–71.

—— (2004) "Now Is the Time." *Philosophical Transactions of the Royal Society B: Biological Sciences* 359 (1444): 731.

—— (2010) "Hope for Tropical Biodiversity through True Bioliteracy." *Biotropica* 42 (5): 540–42.

Kim, KeChung and Loren B. Byrne (2006) "Biodiversity Loss and the Taxonomic Bottleneck: Emerging Biodiversity Science." *Ecological Research* 21 (6): 794–810.

Kress, W. John, David L. Erickson, F. Andrew Jones, Nathan G. Swenson, Rolando Perez, Oris Sanjur *et al.* (2009) "Plant DNA Barcodes and a Community Phylogeny of a Tropical Forest Dynamics Plot in Panama." *Proceedings of the National Academy of Sciences* 106 (44): 18621–26.

Miller, Scott E. (2007) "DNA Barcoding and the Renaissance of Taxonomy." *Proceedings of the National Academy of Sciences* 104 (12): 4775–76.

Oliver, Ian and Andrew J. Beattie (1993) "A Possible Method for the Rapid Assessment of Biodiversity." *Conservation Biology* 7 (3): 562–68.

Packer, Laurence, J. Gibbs, C. Sheffield, and R. Hanner (2009) "DNA Barcoding and the Mediocrity of Morphology." *Molecular Ecology Resources* 9: 42–50.

Prendini, Lorenzo (2005) "Comment on 'Identifying Spiders through DNA Barcodes'." *Canadian Journal of Zoology* 83 (3): 498–504.

Ratnasingham, Sujeevan and Paul D. N. Hebert (2007) "BOLD: The Barcode of Life Data System (www.barcodinglife.org)." *Molecular Ecology Notes* 7 (3): 355–64.

—— (2013) "A DNA-Based Registry for All Animal Species: The Barcode Index Number (BIN) System." *PLoS ONE.* doi: 10.1371/journal.pone.0066213.

Roberts, Leslie (2001) "Controversial from the Start." *Science* 291 (5507): 1182–88.

Rodman, James E. and Jeannine H. Cody (2003) "The Taxonomic Impediment Overcome: NSF's Partnerships for Enhancing Expertise in Taxonomy (PEET) as a Model." *Systematic Biology* 52 (3): 428–35.

Schindel, David and Scott E. Miller (2005) "DNA Barcoding a Useful Tool for Taxonomists." *Nature* 435 (17) (May 5). doi: 10.1038/435017b.

Schoch, Conrad L., Keith A. Seifert, Sabine Huhndorf, Vincent Robert, John L. Spouge, C. André Levesque *et al.* (2012) "Nuclear Ribosomal Internal Transcribed Spacer (ITS) Region as a Universal DNA Barcode Marker for Fungi." *Proceedings of the National Academy of Sciences* 109 (16): 6241–46.

Seifert, Keith A., Robert A. Samson, Jeremy R. deWaard, Jos Houbraken, C. André Lévesque, Jean-Marc Moncalvo *et al.* (2007) "Prospects for Fungus Identification Using COi DNA Barcodes, with Penicillium as a Test Case." *Proceedings of the National Academy of Sciences* 104 (10): 3901–6.

Smith, M. Alex, Norman E. Woodley, Daniel H. Janzen, Winnie Hallwachs, and Paul D. N. Hebert (2006) "DNA Barcodes Reveal Cryptic Host-Specificity within the Presumed Polyphagous Members of a Genus of Parasitoid Flies

(Diptera: Tachinidae)." *Proceedings of the National Academy of Sciences* 103 (10): 3657–62.

Sokal, Robert R. and Peter H. A. Sneath (1963) *Principles of Numerical Taxonomy.* San Francisco: W. H. Freeman.

Stevens, Mark I., David Porco, Cyrille A. D'Haese, and Louis Deharveng (2011) "Comment on 'Taxonomy and the DNA Barcoding Enterprise' by Ebach (2011)." *Zootaxa* 2838: 85–88.

Tangley, Laura (1990) "Cataloging Costa Rica's Diversity." *BioScience* 40 (9): 633–36.

Taylor, H. R. and W. E. Harris (2012) "An Emergent Science on the Brink of Irrelevance: A Review of the Past 8 Years of DNA Barcoding." *Molecular Ecology Resources* 12 (3): 377–88.

Vences, Miguel, Meike Thomas, Ronald M. Bonett, and David R. Vieites (2005) "Deciphering Amphibian Diversity through DNA Barcoding: Chances and Challenges." *Philosophical Transactions of the Royal Society B: Biological Sciences* 360 (1462): 1859–68.

Vernooy, Ronnie, Ejnavarzala Haribabu, Manuel Ruiz Muller, Joseph Henry Vogel, Paul D. N. Hebert, David E. Schindel *et al.* (2010) "Barcoding Life to Conserve Biological Diversity: Beyond the Taxonomic Imperative." *PLoS Biol* 8 (7): e1000417.

Vialle, Agathe, N. Feau, M. Allaire, M. Didukh, F. Martin, J. M. Moncalvo *et al.* (2009) "Evaluation of Mitochondrial Genes as DNA Barcode for Basidiomycota." *Molecular Ecology Resources* 9: 99–113.

Wheeler, Quentin D., Peter H. Raven, and Edward O. Wilson (2004) "Taxonomy: Impediment or Expedient?" *Science* 303 (5656): 285.

Wilkening, Stefan, Manu M. Tekkedil, Gen Lin, Emilie S. Fritsch, Wu Wei, Julien Gagneur *et al.* (2013) "Genotyping 1000 Yeast Strains by Next-Generation Sequencing." *BMC Genomics* 14 (1): 90.

Will, K. W., B. D. Mishler, and Q. D. Wheeler (2005) "The Perils of DNA Barcoding and the Need for Integrative Taxonomy." *Systematic Biology* 54 (5): 844–51.

Wilson, Edward O. (1985) "The Biodiversity Crisis: A Challenge to Science." *Issues in Science and Technology* 2: 20–29.

——— (2004) "Taxonomy as a Fundamental Discipline." *Philosophical Transactions of the Royal Society B: Biological Sciences* 359 (1444): 739–39.

Wong, Eugene H. K. and Robert H. Hanner (2008) "DNA Barcoding Detects Market Substitution in North American Seafood." *Food Research International* 41 (8): 828–37.

PART III

The structure of evolutionary theory

Darwin's theory and the value of mathematical formalization

R. Paul Thompson

Prolegomena

Charles Darwin's intellectual genius and clear, elegant writing style pervade the impressive corpus of his work; never more so than in *The Origin of Species*. The evidence he marshalled in support of biological evolution is impressive, as is the theoretical framework he developed to mechanistically explain the evolution of the vast array of living things, the traits they possess, and the adaptation of those traits to an environment. Nonetheless, Darwin had an intellectual weakness: he had little talent for, or training in, mathematics. There is not a single equation in the *Origin*.[1]

I have in many places argued that mathematics is the language of science, following Bradwardine's and Galileo's salient expression of this position:

> [I]t is [mathematics] which reveals every genuine truth, for it knows every hidden secret, and bears the key to every subtlety of letters; whoever, then, has the effrontery to study physics while neglecting mathematics, should know from the start that he will never make his entry through the portals of wisdom. (Thomas Bradwardine, *Tractatus de continuo*,[2] *c.* 1330s)
>
> Philosophy is written in this grand book, the universe, which stands continually open to our gaze. But the book cannot be understood unless one first learns to comprehend the language and read the letters in which

---

This chapter connects with works of Michael Ruse in three ways. First, it builds on his contention, in his 1973 book *Philosophy of Biology*, that evolutionary theory is as logically and epistemologically deep and rich as the theories of physics. Second, it employs his insight, in 1975, that Darwin owed a debt to philosophy, in particular to Whewell's and Herschel's philosophies of science (Ruse 1975). And, as an aside, speaking of intellectual debts, it's hard to overestimate the debt philosophy of biology owes to Michael. Third, he pioneered, in his *The Darwinian Revolution: Science Red in Tooth and Claw* (1979), an integrated approach in history and philosophy of biology that demonstrated the interdependence of history and philosophy of biology. This chapter attempts to emulate that integrated approach.

I am grateful for the very insightful comments of Jean Gayon and Denis Walsh.

[1] There are tabulations and simple ratios in what he called his "bigger book" – the one from which the *Origin* was an abstract – but nothing more (see Stauffer's 1987 editing of the first volume of the "bigger book," especially Darwin's chapter "Variation under Nature").

[2] As quoted in Weisheipl 1967, 94.

it is composed. It is written in the language of mathematics, and its characters are triangles, circles and other geometric figures without which it is humanly impossible to understand a single word of it; without these, one wanders about in a dark labyrinth. (Galileo Galilei, *Il saggiatori*, 1623[3])

Darwin, it seems, stands as a striking exception to my claim. That, however, is not the case. My central claim in this chapter is that until his theory was formalized in the language of mathematics, its implications, predictions, and use in explanations was far from fully realized. Indeed, the confusion and controversies in the six decades following the publication of the *Origin* illustrate and underscore this point. In making the case, I examine three stages. First, I set out the theory as Darwin structured and expounded it in the *Origin*, giving it the axiomatic, albeit informal, rendition that I contend Darwin intended. Second, I examine the early attempts to employ mathematics. The focus is on the underdeveloped elements in Darwin's exposition: heredity and variation. The main characters are Galton, Pearson, Mendel, Weldon, and Bateson; the first three bring mathematics fully to bear on these matters. Finally, I look at the developments from 1910 to 1930; by 1930 Darwin's theory had been given a rigorous mathematical formalization. The mathematical formalizations of Fisher, Haldane, Wright, and Weinberg[4] map neatly on to Darwin's formulation. I use Fisher's 1930 *The Genetical Theory of Natural Selection* to demonstrate this point, mostly because it was the first holistic account that brought together Darwinian evolution, Mendelian genetics, and biometry. Although there are important differences between Fisher, Haldane, Wright, and Weinberg,[5] the case I make using Fisher can easily be made using any of the others.

Fisher's formalization is not entirely congruent with Darwin's informal theory; Fisher focuses on the population expansion part of Darwin's and Wallace's "struggle" but not on the means of subsistence constraint. For Fisher, Darwin's "Malthusian struggle" relates to population dynamics, in particular reproductive value, v,[6] and its change over time, dv. It is

[3] Commonly known in English as *The Assayer* (Middle English "assay" assimilated to French "essayer"). A superb translation by Stilman Drake can be found in Galileo 1623.
[4] Hardy obviously played an important role in the development of the mathematical formalization but it was a single incursion, whereas the others in this list engaged in sustained work on the formalization. It is worth noting that Fisher never mentions Hardy or Weinberg in *The Genetical Theory of Natural Selection* (1930).
[5] These differences are clearly set out in Provine 1971 and Gayon 1998.
[6] Crow (2002) claims, correctly I think, that this was Fisher's central conceptual contribution: "Although the Malthusian parameter was not original with Fisher, he introduced a deeper concept that was strikingly new, reproductive value" (1314).

that aspect on which he focuses and uses to generate natural selection and ultimately the variable, W,[7] that represents it in the calculus.[8]

The structure of Darwin's theory in the *Origin*

Darwin modeled his theory on the philosophies of science of John F. W. Herschel (especially his 1831 *A Preliminary Discourse on the Study of Natural Philosophy*) and William Whewell (especially his 1840 *Philosophy of the Inductive Sciences*) (see Ruse 1975). Whewell seems to have had the greater influence on Darwin. For both, Newtonian Mechanics stood as the exemplar of a scientific theory; from a few exceptionally abstract generalizations – the three laws of motion and the law of gravitational attraction – all other generalizations (laws) can be deduced. Their only justification is that the less general statements that are deduced from them accord with the observed behavior of things in the world. Theories understood this way integrate a large body of knowledge and, as a result, provide robust explanations and predictions. So, schematically, a theory for both Herschel and Whewell would appear as described in Figure 6.1.

  With this understanding of the structure of scientific theories in mind, Darwin, in a very revealing passage, captures two features of his theory. First, that "the conditions of existence" is a law and is embraced by natural selection. Second, that the law of "the unity of type" is subsumed under (derivable from) the law of "the conditions of existence." He, thereby, demonstrates that there are general laws from which other laws can be derived.

> It is generally acknowledged that all organic beings have been formed on two great laws – Unity of Type, and the Conditions of Existence. By unity of type is meant that fundamental agreement in structure, which we see in organic beings of the same class, and which is quite independent of their habits of life. On my theory, unity of type is explained by unity of descent. The expression of conditions of existence, so often insisted on by the illustrious Cuvier, is fully embraced by the principle of natural selection. For natural selection acts by either now adapting the varying parts of each being to its organic and inorganic conditions of life; or by having adapted them during long-past periods of time: the adaptations being aided in some cases by use and disuse, being slightly affected by the direct action of the external

---

[7] To be more precise, W is the genetic variance of fitness, which equals the rate of increase of fitness (M) in the population or as Fisher also states "the actual rate of increase of fitness" (1930, 42).

[8] I am grateful for an exchange with Denis Walsh that disclosed a need to be explicit about this difference between Darwin and Fisher. This difference is important in many contexts, especially as more variables were added to the calculus over the next 80-odd years, such as a proportionality variable for frequency-dependent selection and a coefficient of relatedness for kin selection. It is not, however, a difference that matters to my thesis.

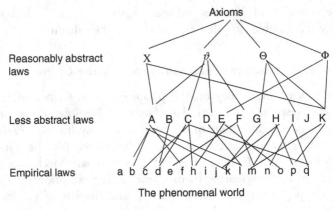

Figure 6.1 Theories

conditions of life, and being in all cases subjected to the several laws of growth. Hence, in fact, the law of the Conditions of Existence is the higher law; as it includes, through the inheritance of former adaptations, that of Unity of Type. (Darwin 1859, 165)

In another succinct passage, he bundles three of his general laws together:

> Finally, it may not be a logical deduction, but to my imagination it is far more satisfactory to look at such instincts as the young cuckoo ejecting its foster-brothers, – ants making slaves, – the larvae of ichneumonidae feeding within the live bodies of caterpillars, – not as specially endowed or created instincts, *but as small consequences of one general law, leading to the advancement of all organic beings, namely, multiply* [reproduction and heredity], *vary* [variation], *let the strongest live and the weakest die* [struggle for existence]. (Darwin 1859, 193, emphases added)

Although, in this passage, he states these three aspects as a single law, throughout the *Origin* he deals with them separately as is appropriate given their independence. Darwin considered these to be the most general laws of his theory (what today would be called axioms, although Darwin never used that term). To these, he adds a fourth, the artificial demarcation of species:

> Finally, then, varieties have the same general characters as species, for they cannot be distinguished from species, except, firstly, by the discovery of intermediate linking forms, and the occurrence of such links cannot affect the actual characters of the forms which they connect; and except, secondly, by a certain amount of difference, for two forms, if differing very little, are generally ranked as varieties, notwithstanding that intermediate linking forms have not been discovered; but the amount of difference

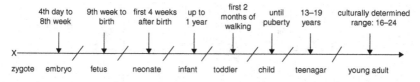

Figure 6.2 Human development

considered necessary to give to two forms the rank of species is quite indef-inite. (Darwin 1859, 58–59)

Certainly no clear line of demarcation has as yet been drawn between species and sub-species that is, the forms which in the opinion of some naturalists come very near to, but do not quite arrive at the rank of species; or, again, between sub-species and well-marked varieties, or between lesser varieties and individual differences. These differences blend into each other in an insensible series; and a series impresses the mind with the idea of an actual passage. (Darwin 1859, 51)

Species, sub-species, and varieties are a human imposition on an insen-sible continuous series. This is analogous to embryological development. Human development from a fertilized egg to an adult is a continuous series but those who study development carve it up into stages. This aids scientific discussion, providing a vocabulary to allow parts of the series to be studied and described but they are nonetheless arbitrary (see Figure 6.2).

As from these remarks it will be seen that I look at the term species, as one arbitrarily given for the sake of convenience to a set of individuals closely resembling each other, and that it does not essentially differ from the term variety, which is given to less distinct and more fluctuating forms. (Darwin 1859, 521)

This is an important point. If species are something "real" in nature, akin to a natural kind, then transmutation will not be possible; species will not be mutable. If species are immutable, no matter how many trait changes occur, a new species cannot emerge and, hence, evolution is impossible.

Pulling this together and using more contemporary language, Darwin posited and argued for these four axioms:[9]

Axiom 1: The demarcation of species is artificial (organisms manifest an insensible gradation of forms).

---

[9] Jean Gayon (2008) provides a reconstruction along the same lines, except that he identifies postu-lates and is principally interested in a contemporary issue arising from Stephen J. Gould's paper, "Is a New and General Theory of Evolution Emerging?" (1980)

R. Paul Thompson

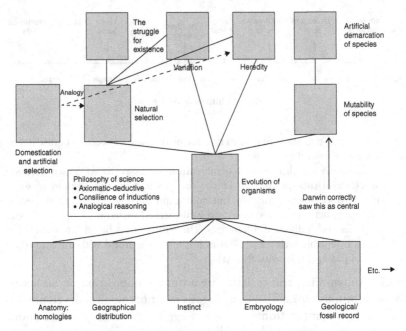

Figure 6.3 Darwin's axioms and the consilience of inductions

Axiom 2: There is a struggle for existence among organisms (the conditions of existence).

Axiom 3: There exist small trait differences among organisms (variation).

Axiom 4: Small trait differences are heritable (heredity).

With these four axioms, his theory, as he sets it out in the *Origin*, can be represented schematically (see Figure 6.3).

Darwin provides compelling narratives and a host of evidence for this "theory" and some elements of it; nonetheless, the deductions are informal and far from rigorous. The internal connectivity and the relationship of the theory to phenomena is, consequently, also informal.

Apart from the overall axiomatic sketch, two other features echo Whewell's philosophy of science: the use of analogy to connect known causal events with hidden causal events (artificial selection and its results with natural selection and heredity) and the consilience of inductions. Darwin knew that the more domains of natural history (biology) his theory could explain, the more robust it would be. Darwin's initial motivation was to understand the observed gradations in organisms.

The most curious fact is the perfect gradation in the size of the beaks in the different species of Geospiza, from one as large as that of a hawfinch to that of a chaffinch, and (if Mr. Gould is right in his including his sub-group, Certhidea, in the main group), even to that of a warbler ... Seeing this gradation and diversity of structure in one small, intimately related group of birds, one might really fancy that from an original paucity of birds in this archipelago, one species had been taken and modified for different ends. (Darwin 1839, 364–65)

But the theory that explains this gradation also explains the biogeographic distribution of organisms, anatomical features of organisms such as homologies – similarities in structure and rudimentary organs, for example – embryology, instinctive behavior, the fossil record, what fossils are there, why there are gaps in the sequences and the temporal ordering of fossils, and so on. Hence, Whewell's consilience of induction pervades this theory.

The ink was barely dry on the *Origin* before the central mechanism of Darwin (and Wallace) – natural selection acting on small individual variations – was contested, even by evolutionists who accepted that natural selection is a key mechanism. Thomas Huxley, an ardent evolutionist who accepted natural selection, immediately voiced this objection: "The only objections that have occurred to me are, 1st that you have loaded yourself with an unnecessary difficulty in adopting *Natura non facit saltum* so unreservedly" (letter to Darwin dated November 23, 1859: Darwin 1959, vol. II, 27). Given the centrality of this view to Darwin's theory, it stands as an appropriate test case of my claim that the lack of a mathematical formalization lies at the heart of the explanation of this debate and others.

Continuous versus discontinuous evolution

In the years following the publication of the *Origin*, six people figure prominently in the quest to understand heredity, variation and the material basis of natural selection: Francis Galton, W. F. R. Weldon, Karl Pearson, William Bateson, Wilhelm Johanssen, and August Weismann. In the last decade of the nineteenth century two camps had emerged and feuding had begun (a feud, it is worth acknowledging, based as much on personalities as conceptual differences). In another ironic twist, Francis Galton was viewed by both camps as a founder of their position. More importantly, Galton challenged Darwin's theory of heredity and developed his own – one that came very close to Mendel's.

The theory of heredity that Darwin adopted by 1868 was named pangenesis. According to this theory every organ of the body sheds small particles (gemmules),[10] which collect in the organs of reproduction and through which they are passed to the offspring. Adapting Lamarck's "inheritance of acquired characteristics," Darwin reasoned that since the environment could cause changes in the organs of the body, it could cause changes in the gemmules that an organ shed. This was the source of new variation in every generation. The details of how this mechanism of pangenesis functioned were sketchy. Darwin saw this mechanism as consistent with his view that heredity resulted in characteristics blending from generation to generation (blending heredity). Without blending, he held, the creation of abundant variation in every trait from generation to generation would render the persistence of any trait impossible; trait persistence is essential to evolution. So, blending heredity tames the relentless creation of variability. Galton rejected pangenesis and attempted to demonstrate experimentally that it was flawed, but with each experimental challenge, Darwin cleverly altered the relevant aspect of the theory. It is Galton's mathematical work, more than his experimental work, on heredity that matters most to my thesis.

The postulate of Darwin's pangenesis that Galton found entirely untenable was the inheritance of acquired characteristics. Two principle issues were: the non-effect of amputations on the next generation and reversions, where a trait of an ancestor reappears several generations later. The former Darwin explained away with the claim that one would have expected there to be environmental influences that are unable to change the gemmules. Rendering his view immune to Weissman's experiments involving severing the tails of mice for many generations and discovering no effect on the tails of offspring. Darwin attributed reversion to dormant gemmules:

> These granules may be called gemmules. They are collected from all parts of the system to constitute the sexual elements, and their development in the next generation forms a new being; but they are like-wise capable of transmission in a dormant state to future generations and may then be developed. (Darwin 1868, vol. II, 370)

Galton was, understandably, unimpressed and set out to develop an alternative. His first exposition of his alternative theory was published in *Proceedings of the Royal Society* in 1872 under the title, "On Blood-Relationship." Unfortunately, it was under-developed and verged on incoherence. In 1875, Galton produced a popular, and hence more accessible,

---

[10] Darwin considered the pangenes to be "autonomous elements of the body," perhaps "cells" (Darwin 1868, vol. II, 370). He tentatively, in this volume, considered his "gemmules" as "cell-gemmules" (374).

account of his theory in the *Contemporary Review* under the title "A Theory of Heredity." He introduced a new term, stirp, which designates, "the sum total of the germs, gemmules, or whatever they be called, which are to be found, according to every theory of organic units, in the newly fertilised ovum"; he prefers "germs" to "gemmules," mostly, it seems, to distinguish his understanding of the hereditary units from Darwin's. The ovum contains the stirp as well as required nutrients; hence, the stirp in contemporary terms is similar to a nucleus in a cell.

With this concept of a stirp in place, he set out four postulates, which he claims are uncontroversial:

> We will begin with a statement of the four postulates that seem to be almost necessarily implied by any hypothesis of organic units, and which are included in that of Pangenesis. The first is, that each of the enormous number of quasi-independent units of which the body consists, has a separate origin, or germ. The second is, that the stirp contains a host of germs, much greater in number and variety than the organic units of the structure that is about to be derived from them; so that comparatively few germs achieve development. Thirdly, the germs that are not developed, retain their vitality; they propagate themselves while still in a latent state, and they contribute to form the stirps of the offspring. Fourthly, organisation wholly depends on the natural affinities and repulsions of the separate germs; first in their stirpal, and subsequently during all the processes of development.
>
> Proofs of the reasonableness of these postulates are especially to be found in the arguments of Mr. Darwin: that there is at least fair ground to believe in their reasonableness, may be shown in a cursory manner. (Galton 1876, 331)

Which Galton then proceeds to do. So, there is a separate germ for every unit of the body; one can assume these are cells though Galton continues to use the vague term units, which some have assumed were, in part, collections of cells such as organs. There are more germs than required for determining the structure of the adult organism; that is, not all the germs in a stirp are expressed in any specific organism. Those germs that do not participate in the development of an organism retain their potential for doing so in the development of another organism – a sibling or offspring. Construction of the adult organism is a dynamical one with germs attracting and repulsing each other.

These postulates form the foundation of his theory but, as Galton was aware, more is required. Principally, some method of reduction is required. If germs from a male and female combine to create a stirp, there will be double the number in the newly formed stirp than in either parent. In just a few generations the number of germs in a stirp would have ballooned. In true Darwinian fashion, he thought the reduction resulted from a struggle for existence, which half the germs would lose.

Darwin would find little to which to object in the theory to this point; the point of divergence, presented in the second half of the paper, centered on how the germs in a stirp behave and how stirps arise:

> The conclusion from what has thus far been said is amply confirmed by observation; it is: – (1.) That the contents of the stirp must segregate into septs, or divisions, and these septs must subdivide again and again, just as a large political party may repeatedly subdivide itself into different factions. (2.) That the dominant germs in each successive sept are those that achieve development. (3.) That it is the residual germs and their progeny that form the sexual elements or buds.
>
> ...
>
> This is a very different supposition to that of the free circulation of gemmules in Pangenesis, yet it seems to have the merits of that theory (so far as the group of cases are concerned which we are now considering, namely, the inheritance of qualities that were congenital in the ancestry), and at the same time to be free from the many objections that urged against it. (Galton 1876, 340–41)

Here Galton, inelegantly, veers in the direction of the still unknown theory Mendel had ten years earlier, elegantly, set out. Contrary to Darwin, Galton separated the sexual elements from the units making up the rest of the organism. The fertilized ovum divides over and over again resulting in the organism and so do the germs (the contents of the stirp). Dominant germs result in the various units of the organism. Residual germs from all these divisions are the sexual elements. Galton provided little clarification of the details about "residual germs." The implication of this view of the separation of sexual elements from other units of the body was that changes to those other units did not affect the germs in sperm and ovum and, hence, acquired characteristics could not be inherited.

Galton also focused his attention on traits of organisms which quantitatively vary – height, weight, number of body hairs, speed of running, for example – and, most significantly, conformed to a normal distribution. Here Galton was applying to organisms what he had so successfully employed in his other studies and, by so doing, introduced a statistical approach to traits of organisms. Underlying this variation of each trait were the germs in the stirp. Traits were frequently the result of many germs working in concert and could be affected by the environment; nutritional deficiencies could result in underdeveloped height, for example. Here, he anticipated the contemporary field of quantitative genetics, a sophisticated extension of Mendelian (population) genetics, which takes into account that most traits – especially in mammals – result from the interaction of many genes, can be influenced by the environment, and quantitatively vary in exactly the way Galton described.

George Darwin, Charles Darwin's second son, excelled in mathematics at Cambridge; in 1868 he was second wrangler (i.e. second among those awarded a first-class degree in the three mathematical examinations). Darwin turned to George for assistance when mathematical presentations taxed his ability. Galton's statistical presentation of features of his theory of heredity was just such a case. It eluded Darwin, so he turned to his son for assistance. Based on his correspondence with George, Darwin wrote to Galton on December 18, 1875:

> MY DEAR GALTON,
> George has been explaining our differences. I have admitted in new Edit, (before seeing your essay) that perhaps the gemmules are largely multiplied in the reproductive organs; but this does not make me doubt that each unit of the whole system also sends forth its gemmules. You will no doubt have thought of the following objection to your view, and I should like to hear what your answer is. If 2 plants are crossed, it is often or rather generally happens that every part of the stem, leaf – even to the hairs – and flowers of the hybrid are intermediate in character; and this hybrid will produce by buds millions on millions of other buds all exactly reproducing the intermediate character. I cannot doubt that every unit of the hybrid is hybridised and sends forth hybridised gemmules. Here we have nothing to do with the reproductive organs. There can hardly be a doubt, from what we know, that the same thing would occur with all those animals which are capable of budding and some of those (as the compound Ascidians) are sufficiently complex and highly organised.
> Yours very sincerely, CH. DARWIN. (Pearson 1924, 189)

On December 19, Galton responded:

> MY DEAR DARWIN, The explanation of what you propose does not seem to me in any way different on my theory, to what it would be in any theory of organic units. It would be this:
> Let us deal with a single quality, for clearness of explanation, and suppose that in some particular plant or animal and in the same particular structure, the hybrid between white and black forms was exactly the intermediate, viz: grey – thenceforward for ever. Then a bit of the tinted structure under the microscope would have a form which might be drawn as in a diagram, as follows: –
> ...
> In (1) we see that each cell had been an organic unit (quoad colour). In other words the structural unit is identical with the organic unit.
> In (2) the structural unit would not be an organic unit but it would be an organic *molecule*. It would have been due to the development, not

White form.   Black form.

(1)   (2)

of one gemmule but of a group of gemmules, in which the black and the white species would, on statistical grounds, be equally numerous (as by the hypothesis, they were equipotent).

The larger the number of gemmules in each organic molecule, the more *uniform* will be the tint of greyish in the different units of the structure. It has been an old idea of mine, not yet discarded and not yet worked out, that the number of the units in each molecule may admit of being discovered by noting the relative number of cases of each grade of deviation from the mean greyness. If there were two gemmules only, each of which might be either white or black, then in a large number of cases one-quarter would always be quite white, one-quarter quite black, and one-half would be grey. If there were three molecules, we should have 4 grades of colour (1 quite white, 3 light grey, 3 dark grey, 1 quite black and so on according to the successive lines of "Pascal's triangle"). This way of looking at the matter would perhaps show (a) whether the number in each given species of molecule is constant, and (b), is so, what those numbers were. Ever very faithfully yours, FRANCIS GALTON. (Pearson 1924, 189–90)

Karl Pearson remarked, "This letter shows how very closely Galton's thoughts at this time run on Mendelian lines" (Pearson 1924, 190).

Pascal's triangle (see Figure 6.4) is a remarkable pattern of numbers.[11] The number in each row, after row 0, is the sum of the two numbers above it. So, in row 1, the number above each of the two units is 1; in row 2, the numbers

---

[11] The mathematician and philosopher Blaise Pascal is credited with the triangle, largely because he most clearly drew out its fascinating properties. It dates back to at least the middle of the twelfth century in the writing of the Chinese mathematician Yanghui and is found in Persian mathematical texts.

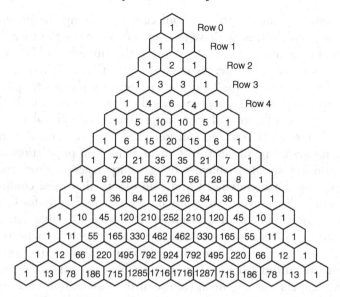

Figure 6.4 The numerical array known as Pascal's triangle

above 2 are 1 and 1 (1 + 1 = 2); in row 3, the number above each of the 3s are 1 and 2 (1 + 2 = 3); in row 4, the numbers above the 4s are 1 and 3 (1 + 3 =4) and above the 6 are 3 and 3 (3 + 3 =6); and so it continues. Down the sides of the triangle are only 1s because only the single number 1 is above them. The structure of this pattern embodies many other mathematical patterns such as prime numbers, Fibonacci's sequence, Sierpinski's fractal, and polygonal numbers.

For Galton, it was the fact that it embeds the binomial coefficients. A binomial has two values, a and b, for example, which are added $(a + b)$. A simple example is a coin toss, where a = heads and b = tails. Each toss is another $(a + b)$. Consequently, two tosses is $(a + b)(a + b)$ or $(a + b)^2 = aa + ab + ba + bb = a^2 + 2ab + b^2$; which are the values in row 2 of Pascal's triangle (1, 2, 1). The expansion of $(a + b)^3$ is: $a3 + 3a2b + 3ab2 + b3$; the co-efficients are the same as row 3 of Pascal's triangle (1, 3, 3, 1). $(a + b)^4 = a^4 + 4a^3b + 6a^2b^2 + 4ab^3 + b^4$ (row 4 of Pascal's triangle).

When Galton suggested that, "If there were three molecules, we should have 4 grades of colour (1 quite white, 3 light grey, 3 dark grey, 1 quite black and so on according to the successive lines of 'Pascal's triangle')", he was assuming black and white $(b + w)$ gemmules[12] and raising the combination

---

[12] In this letter he uses Darwin's term not his own, "germ."

to the power of the number of molecules containing the gemmules. Hence, with 3 molecules, we have $(b + w)^3 = b3 + 3b2w + 3bw2 + w3$ (that is grades of color in the proportion: 1:3:3:1 = the multiples of his "4 grades of colour" = the fractional representation: 1/8th, 3/8th, 3/8th, 1/8th. With 4 molecules, we have: $(b + w)^4 = b^4 + 4b^3w + 6b^2w^2 + 4bw^3 + w^4$: 5 grades of color in the proportion 1:4:6:4:1. So Pascal's triangle gives easily both the number of grades of color and the ratio/proportion of each color.

Galton's use of binomial expansions had a more general purpose. He was interested in the distribution of traits in a population: color, height, running speed, problem-solving ability, reaction time, and the like. His studies had convinced him, and others, that these conformed to a Gaussian distribution (the normal curve). Importantly for Galton's theory, as the exponent of a binomial increases, the binomial curve approaches a Gaussian curve. When the binomial exponent reaches 1,000, the curves are hard to distinguish. Hence, Galton has now connected his theory of heredity – a theory which mathematically comes close to Mendel's – to a general feature of the distribution of traits in a population. What he did not do was connect his theory to Darwin's theory of evolution; Galton continued to accept discontinuous evolution. Hence, he developed a mathematically sophisticated theory of heredity but he failed to find in it the resolution of a contentious issue in the theory of evolution.

In 1900 Mendel's theory entered the fray. His work and theory centered on three things: first, he provided results from a number of well-designed, expertly conducted and recorded experiments on hybridization; second, he provided a mathematical description of the results in terms of ratios; and third, he provided a theory to account for the ratios. That theory postulated entities (factors – today known as alleles), properties of the entities (e.g. they can be dominant or recessive), and a dynamical system in which the entities exist in pairs in all cells, give rise, as pairs, to the characteristics of the organism, segregate in the reproductive organs (the needed reduction Galton identified as necessary and Weismann had observed in a somewhat different context), and recombine in fertilization to create new pairs that determine the characteristics of the offspring.

William Bateson embraced Mendel's theory soon after it became widely known in 1900. He regarded it as incompatible with Darwinian evolution in three respects: it was incompatible with natural selection, with blending heredity, and with continuous evolution. Bateson also contended that is was incompatible with Weldon's and Pearson's biometrical approach.

Obviously, no unified theory of evolution was going to emerge from Bateson's Mendelism.

The controversy over which general view was correct had become acrimonious well before 1900 and continued after Mendel's theory became known. Just how poisoned the atmosphere had become by 1902 can be seen in G. Udny Yule's[13] comments on Bateson and Saunders's "Reports to the Evolution Committee of the Royal Society" and Bateson's *Mendel's Principles of Heredity: A Defence*.

> The sections of the two volumes which do appear to call for criticism and review are those relating to the bearing of Mendel's results on the conceptions of heredity in general, and on the work of Mr. Francis Galton and Professor Pearson in particular. Mr. Bateson devotes many words to these questions, but one cannot help feeling that his speculations would have had more value had he kept his emotions under better control; the style and method of the religious revivalist are ill-suited to scientific controversy. It is difficult to speak with patience either of the turgid and bombastic preface to "Mendel's Principles," with its reference to Scribes and Pharisees, and its Carlylean inversions of sentence, or of the grossly and gratuitously offensive reply to Professor Weldon and almost equally offensive adulation of Mr Galton and Professor Pearson. A writer who indulges himself in displays of this kind loses his right to be treated either as an impartial critic or a sober speculator. Mr. Bateson is welcome to dissent from Professor Weldon's opinions, but it would have been well if he had imitated the studied moderation and courtesy of his article. (Yule 1902, 194)

This acrimonious environment blinkered almost all of the participants; the focus of Weldon and Pearson, who agreed with Darwin on continuous evolution, was turned away from Mendel by Bateson's interpretation and strident personality. In this environment, Punnett, a committed Mendelian, gave a lecture to the Epidemiological section of the Royal Society of Medicine on February 28, 1908 titled "Mendelism in Relation to Disease" (see Punnett 1908). During the discussion that followed, Yule expressed that he had less optimism that Mendelism would yield the benefits in medicine that Punnett had claimed. Yule in his comments was not rejecting Mendelism, although a few commentators have cast his remarks that way. Most of his objections to Punnett's paper were focused on whether, in this or that case, more could be gained by a biometric (accounting) approach to a disease than a Mendelian one, even assuming Mendel to be correct with respect to the germinal description:

[13] Yule was one of a very few scholars to claim that Darwinism, Mendelism, and biometrics were compatible.

After all, what had to be dealt with was the character which was exhibited, and he agreed with the last speaker that in cases like these the actuarial method was likely to yield more valuable information to the medical man than a discussion on the basis of germinal laws, which might hold for the germ-cells but need not hold for the body, seeing how much the element of circumstance entered into the matter. Other factors as important as heredity must be taken into consideration. The actuarial statement included what the germinal statement did not, namely, those factors of disturbance which were of equal importance with the factors of pure heredity. (Punnett 1908, 165)

In the course of his comment Yule asserted:

The same applied to the examples of brachydactyly [abnormally short fingers and toes]. The author said that brachydactyly was dominant. In the course of time one would then expect, in the absence of counteracting factors, to get three brachydactylous persons to one normal, but that was not so. There must be other disturbing factors of equal importance. (Punnett 1908, 165)

Punnett was sure Yule was wrong, so asked his friend G. H. Hardy about it. Hardy, a brilliant mathematician – perhaps the best of his generation – quickly produced a demonstration using variables where Yule seems to have used specific frequencies of Mendelian factors. Hardy proved that after the first generation, allelic frequencies would remain the same for all subsequent generations (variation would be preserved) unless something caused a change, such as selection. The general form of this equilibrium is:[14]

$$p^2AA:2pqAB:q^2BB.$$

This statement and demonstration of the equilibrium principle should have resolved the debate about variation;[15] unless specific variation-reducing or

[14] Today it is known as the Hardy–Weinberg Equilibrium in recognition that the German biologist Wilhelm Weinberg independently also published the demonstration of this equilibrium in 1908.
[15] Few know that Karl Pearson had stumbled on the same result in 1904, as this passage from his "Mathematical Contributions to the Theory of Evolution. XII. On a Generalised Theory of Alternative Inheritance, with Special Reference to Mendel's Laws" shows:

Hence, by the above proposition, the distribution of offspring of parents of two couplets is
$4 \times 4 \times 4. \ (\frac{1}{4}u + \frac{2}{4}v + \frac{1}{4}w) \times 4 \times 4 \times 4. \ (\frac{1}{4}u + \frac{2}{4}v + \frac{1}{4}w)$
$= 4^2 \times 4^2 \times 4^2. \ (\frac{1}{4}u + \frac{2}{4}v + \frac{1}{4}w)^2$
and, by induction, the distribution of offspring for the random mating of parents of n couplets is
$4^n \times 4^n \times 4^n. \ (\frac{1}{4}u + \frac{2}{4}v + \frac{1}{4}w)^n.$
This, except for the constant factor $4^n \times 4^n$, is absolutely identical with the distribution of the parental population, and **accordingly if the next generation also mates at random, the mixed race will continue to reproduce itself without change. We therefore reach the following result: –**

variation-increasing factors are present, variation is preserved in a Mendelian system; preservation with no increase or decrease is the steady state.

Although there was considerable experimental work that was considered relevant to the continuous versus discontinuous debate during this period, it was far from decisive[16] and was based on numerous disconnected and antithetical theoretical commitments.[17]

Fisher and the formalization of Darwinian evolution

The title of Fisher's 1930 book, *The Genetical Theory of Natural Selection*, announces its central thesis; genetics (Mendelian genetics) and Darwinian natural selection are compatible and he will provide a unified theory.[18]

Fisher quotes T. H. Huxley at the beginning of his second chapter: "In the first place it is said – and I take this point first, because imputation is too frequently admitted by Physiologists themselves – that Biology differs from the Physico-chemical and Mathematical sciences in being "inexact." Huxley, in the work from which this comes, vigorously denies that biology

> *However many couplets we suppose the character under investigation to depend upon, the offspring of the hybrids – or the segregating generation – if they breed at random* inter se, *will not segregate further, but continue to reproduce themselves in the same proportions as a stable population.*
>
> It is thus clear that the apparent want of stability in a Mendelian population, the continued segregation and ultimate disappearance of the heterozygotes, is solely a result of self-fertilisation; with random cross fertilisation there is no disappearance of any class whatever in the offspring of the hybrids, but each class continues to be reproduced in the same proportions. [Italics are original; **bold** is my emphasis.]
>
> Pearson's main purpose in this paper was to demonstrate that Mendelian heredity did not result in new variability that would also be heritable. If, like Yule, he had seen the compatibility of Mendelism and biometrics he might have seen the importance of his insight into the equilibrium state; regrettably, he did not. Moreover, he failed to notice that the line of reasoning he provides elsewhere in the paper actually demonstrates the enormous variability that recombination created in a Mendelian context.

[16] Yule, in 1903, in a review of Johannsen's book underscores this for the experiments deemed to support discontinuous evolution, and the reasoning from Nilsson-Ehle experimental results underscores it for the experiments that were seen as supporting continuous evolution. Yule, in his review, first summarizes Johannsen's findings and conclusions, and, after urging his audience to read Johannsen's book for themselves rather than relying solely on his summary, he makes the point that the experimental results do not justify Johannsen's claim that selection is ineffective with a pure line. Like Weldon and Pearson, he pointed out that the effect of selection in the experiment was small but not zero, and if not zero then inexorably over time it might be effective.

[17] Pearson continued to identify flaws in the research and the reasoning but was mostly scorned and/or ignored by biologists. Pearson was a mathematician, and his reasoning was mathematically based; few biologists had any sophistication in mathematics and most focused narrowly on biological experiments, which they saw as the route to discovery in biology.

[18] Margaret Morrison's excellent 2002 article on Pearson and Fisher on Mendelism and biometry is a fitting companion to this chapter. Her interests are different but our points of agreement are many and differences few.

is a less exact science in either methods or results than the physico-chemical sciences. He concedes only one difference, that is between mathematical sciences and both the physico-chemical and biological sciences; a difference of degree rather than kind. In a telling and most interesting passage he claims:

> The Mathematician deals with two properties of objects only, number and extension, and all the inductions he wants have been formed and finished ages ago. He is occupied now with nothing but deduction and verification.
>
> The Biologist deals with a vast number of properties, and his inductions will not be completed, I fear, for ages to come; *but when they are, his science will be as deductive and as exact as Mathematics themselves*. (Fisher 1930, 57, 58, emphasis added)

By quoting this passage from Huxley, Fisher signals that he is going to render an exact scientific account of evolution. His numerous comparisons of the biological formulas he constructs with principal, interest and repayments of loans, with thermodynamics, and the like, reinforces this impression, as do claims like: "A similar convention, appropriate in the sense of bringing *the formal symbolism of the mathematics into harmony with the biological facts,* may be used with respect to the period of gestation" (Fisher 1930, 24). To achieve his goal of unification Fisher had to mathematically formalize the relevant axioms of Darwin's theory.[19] His unification is achieved, for the most part, in the first two chapters (51 pages), although in chapters 5 and 7 he provides important amplifications of some key elements. As indicated in the "Prolegomena" above, Fisher's formalization does not depend on the varied empirical work.

There have been many debates about, and refinements to, Fisher's formalization[20] but two things are beyond dispute. First, the internal connections, and the explanatory and predictive consequences of the theory were rendered clear. Second, the concordance of any empirical finding and the theory were deductively determinable. The controversy over continuous versus discontinuous evolution demonstrated that Darwin's informal theory achieved neither of these virtues.

In chapter 1, Fisher clears away some rubble. The two principal conceptual rotten beams that he decisively clears away were blending heredity (on this

---

[19] It would be anachronistic to suggest that Fisher consciously set out to formalize Darwin's axioms as I have reconstructed them; that, however, was the effect of his mathematical formalization. The axiomatic rendering given in the section in the present chapter on "The structure of Darwin's theory in the *Origin*" is mine, and it is unlikely that Fisher consciously structured it that way. The mapping I provide suggests that he was working with a conception something like that provided in that section.

[20] There have also been many misinterpretations, a considerable number, as Jean Gayon notes, of Fisher's own making; they result from his idiosyncratic use of common terms such as "fitness of any organism."

the Mendelians were correct) and the evolutionary importance and efficacy of mutations (on this the biometricians Weldon and Pearson were correct).

In chapter 2, Fisher provides the bulk of the formalization. Three things required attention: the struggle for existence (resulting in natural selection), heredity, and the nature of variation. Heredity was readily provided by Mendel's theory, which Fisher adopted completely. So, the task before him was to integrate into this theory natural selection, which is deducible from the struggle for survival, and variation. With that the task was done. Recall that Darwin's axioms, as I have expressed them in modern terms, are:

> Axiom 1: The demarcation of species is artificial (organisms manifest an insensible gradation of forms).
> Axiom 2: There is a struggle for existence among organisms (the conditions of existence).
> Axiom 3: There exist small trait differences among organisms (variation).
> Axiom 4: Small trait differences are heritable (heredity).

Axiom 1 is a *sine qua non*. Without it evolution cannot occur. Darwin provided a compelling nominalist-style argument for its acceptance. Most evolutionists on all sides of the debates, thereafter, took this for granted – even Fisher simply assumed it. Fisher concentrated on axioms 2, 3, and 4. Here, I focus on his mathematical integration of these into Mendel's theory. Then I demonstrate, what others have not, how the first axiom can be integrated into Mendel's theory. That done, I show that the debate over "continuous" versus "discontinuous" evolution is easily resolved using the unified Darwinian–Mendelian Theory of Evolution, solidifying my claim that had a such a mathematical formulation been given by Darwin, 50 years of wrangling would not have occurred.

Fisher approaches Darwin's axiom 2 (There is a struggle for existence among organisms [the conditions of existence]) by way of the life table, something used by insurance companies and demographers: "In order to obtain a distinct idea of the application of Natural Selection to all stages of the life-history of an organism, use may be made of the ideas developed in the actuarial study of human mortality" (Fisher 1930, 22). Although euphemistically called life tables, they are death tables, just as death insurance is euphemistically called life insurance. And it is death rates and reproduction rates in which Fisher is interested. He states his goal thus:

> The object of the present chapter is to combine certain ideas derivable from a consideration of the rates of death and reproduction of a population of

organisms, with the concepts of the factorial scheme of inheritance, so as to state the principle of Natural Selection in the form of a rigorous mathematical theorem, by which the rate of improvement of any species of organisms in relation to its environment is determined by its present condition. (Fisher 1930, 22)

So, his goal is constructing a mathematical expression of population expansion (or decline), which leads to a struggle for survival. As in most cases of mathematical modeling, he begins with the various elements of the system: in this case, death rate and reproduction rate. First, separate equations for each are generated, then the combined action is generated. For death rate, if l_x = the number in a population living to age x, the death rate at age x (chance of someone dying at age x) is:

$$\mu_x = -\,(1/l_x\, dl\,dx)\, l_x, \text{ which equals } -\,dl\,dx\,(\log l_x).$$

For the reproduction rate, if b_x = the rate of reproduction at age x, the chance of someone living to reproduce during the interval dx is:

$$\rho_x = l_x b_x dx.$$

This can describe the total expectation of reproduction for an organism at birth by integrating over the lifespan of the individual from birth (0) to death (∞); this captures every age during the organism's lifetime in which it could reproduce:

$$\int_0^\infty l_x b_x dx.$$

The relative rate of increase can be included by adding a "Malthusian" variable m. Then, as Fisher points out, "the number of persons in the infinitesimal age interval dx will be proportional to $e^{-mx}l_x b_x dx$, for of those born only a fraction l_x survive to this age" (Fisher 1930, 26). The aggregate for all ages is:

$$\int_0^\infty e^{-mx}l_x b_x dx.$$

To this one can add a valuation that recognizes that, except in a steady state, every age in the life cycle will have a different valuation of reproduction, v_x. Fisher's all-in equation for reproductive value is:

$$n_x\{(b_x v_0 - \mu_x v_x)\, dx + dv_x\}.$$

From this, by differentiation, one gets:

$$dv_x - \mu_x v_x dx + b_x v_0 dx = m v_x dx.$$

Hence Fisher concludes, "Consequently the rate of increase in the total value of the population is m times its actual total, irrespective of its constitution in respect of age" (Fisher 1930, 30), where m (the Malthusian parameter), as defined already, is the relative rate of increase. This expresses the struggle for survival in terms of population dynamics. Natural selection is a deductive consequence assuming the values of m and v_x are environmentally constrained, as Fisher conceived they would be:

> We may ask, not only about the newly born, but about persons of any chosen age, what is the present value of their future offspring; and if present value is calculated at the rate determined as before, the question has the definite meaning – To what extent will persons of this age, on the average, contribute to the ancestry of future generations? The question is one of some interest, since the direct action of Natural Selection must be proportional to this contribution. (Fisher 1930, 27)

Fisher's examination of the variance in chapters 1 and 2 formalizes Darwin's axiom 3 (There exist small trait differences among organisms [variation]) and axiom 4 (Small trait differences are heritable [heredity]). Axiom 4, of course, is formalized directly in Mendelian genetics. Hence, Fisher is mostly exploring the implications of Mendelian genetics on variance, since it follows directly from Mendelian genetics that individual differences are heritable. In chapter 1, Fisher compares blending inheritance, in which Darwin believed, and Mendelian particulate inheritance with respect to variance. There he provides the key mathematical formula needed for unification:

> It has not been so clearly recognized that particulate inheritance differs from the blending theory in an even more important fact. *There is no inherent tendency for the variability to diminish.* In a population breeding at random in which two alternate genes of any factor, exist in the ratio p to q, the three genotypes will occur in the ratio $p^2:2pq:q^2$, and thus ensure that their characteristics will be represented in fixed proportions of the population, however they may be combined with characteristics determined by other factors, provided that the ratio $p:q$ remains unchanged. (Fisher 1930, 9–10, emphasis added)[21]

As is frequently the case, Fisher provides no details. He took it to be obvious to anyone with a modicum of mathematical ability. For Fisher, like Hardy and Weinberg, to whom he makes no reference, this equilibrium, which he notes is never how things behave in nature, is a simple consequence of the particulate theory of heredity. So, once the particulate nature of heredity is adopted, this equilibrium, which preserves variability, is easily demonstrated. This constitutes another axiom, one

---

[21] Fisher indicates two factors that will change the ratio of p and q: chance survival and selection.

constitutive of Mendelian genetics and evolution. This axiom can be stated succinctly:

> Axiom 5: In a randomly mating population with alleles a_1 and a_2 in the ratio $p:q$ in F_0, the alleles will be distributed $p^2(a_1):2pq(a_1a_2):q^2(a_2)$ in all subsequent generations, F_1-F_∞, unless something disturbs the ratio.

Asserting this as an axiom, as Fisher so treats it,[22] rests on a specific rendering of Mendelian genetics. The more common view, and the one Michael Ruse propounded in his *Philosophy of Biology*, is that this equilibrium is derivable from Mendel's first law (segregation in the gametic phase and recombination in the fertilization phase), in which case it would be Mendel's law that is the axiom and this equilibrium would be a highly general deductive consequence of it. There are, however, many reasons to treat the equilibrium as an axiom.

First, Mendel's law is never invoked within the calculus of population genetics. Hartl and Clark, for example, mention segregation of alleles once (2007, 5), where it is defined. It never shows up again and there is no mention of Mendel's first law in the entire book. In the section (2.2) on the Hardy–Weinberg Principle, they write:

> Before proceeding further, it may be helpful to summarize the assumptions we are making:
>
> - The organism is diploid
> - Reproduction is sexual
> - Generations are nonoverlapping
> - The gene under consideration has two alleles
> - The allele frequencies are identical in males and females
> - Mating is random
> - Population size is very large (in theory, infinite)
> - Migration is negligible
> - Mutation can be ignored
> - Natural selection does not affect the alleles under consideration
>
> Collectively, these assumptions summarize the *Hardy–Weinberg model*. (Hartl and Clark 2007, 48)

Segregation is not among the assumptions of the model and there is no mention of Mendel's first law. In Halliburton (2004), only "segregational

---

[22] Fisher himself mentions segregation only in his Preface (twice) and on p. 9. His main claim on p. 8 is "It thus appears that, apart from dominance and linkage, including sex linkage, all the main characteristics of the Mendelian system flow from the assumption of the particulate inheritance of the simplest character." So it is particulate inheritance that was Mendel's genius and that is all that need be assumed; it is all Fisher, in fact, assumes. After p. 8, neither segregation nor Mendel is mentioned again.

load" is mentioned (376). Mendelian segregation is never mentioned nor is Mendel's first law.

Second, all the perturbing factors known to occur in actual populations, such as selection and meiotic drive, are represented as additional variables in the equilibrium equation. Third, the foundational role of the equilibrium is rendered obvious; it specifies an equilibrium state similar to Newton's first law, a Nash equilibrium in game theory, a supply and demand equilibrium in Locke's economic theory, and so on.

Fourth, neither Hardy nor Weinberg provides a derivation of the equilibrium from Mendel's first law. Hardy (1908) sets up his demonstration this way:

> Suppose that *Aa* is a pair of Mendelian characters, *A* being dominant, and that in any given generation the numbers of pure dominants (*AA*), heterozygotes (*Aa*), and pure recessives (*aa*) are as p:2q:r. Finally, suppose that the numbers are fairly large, so that the mating may be regarded as random, that the sexes are evenly distributed among the three varieties, and that all are equally fertile.

There is no mention of segregation in his paper and no mention of Mendel's first law, only the suppositions just given. He then claims:

> **A little mathematics of the multiplication-table type** is enough to show that in the next generation the numbers will be as
>
> $$(p+ q)^2 : 2(p+ q)(q+r) : (q+r)^2,$$
>
> or as $p_1:2q_1:r_1$, say. The interesting question is – in what circumstances will this distribution be the same as that in the generation before? It is easy to see that the condition for this is $q^2 = pr$. And since $q_1^2 = p_1r_1$, whatever the values of p, q and r may be, the distribution will in any case continue unchanged after the second generation. (Hardy 1908, 49, emphasis added)

Finally treating the equilibrium (H–W) as an axiom provides a cleaner distinction between the ontology and the dynamics of the theory. Segregation becomes a property of alleles (part of the definition) in the same way that Newtonian bodies have the properties of separate identities and spatial location; Newton considers this part of what it is to be a body. The H–W equilibrium provides part of the dynamical description of how populations change, or not, over time. It describes formally how the system behaves over time with no perturbing factors in the same way Newton's first law provides a description of the dynamics of the system where there are no perturbing factors (forces in his case);

in both cases, other dynamics relating to perturbing factors then come
into play.[23]

This equilibrium, along with the struggle and its result, natural selection, is all Fisher needs to unify Darwin's and Mendel's theories. In the
section on variance in chapter 2, he is exploring a different, although
closely related, issue, one that brings biometry into the fold: "Let us
now consider the manner in which any quantitative individual measurement, such as human stature, may depend upon the individual genetic
constitution" (Fisher 1930, 30). In this section, he provides a mathematical (statistical) treatment of ratios and the effect of gene substitutions for
quantitative (biometric) traits such as height.[24] Height was prominent in
the biometric analyses of Galton and his followers, so it is not an accident
that Fisher explores the case, signaling thereby the connection of biometry
to his unification.

Fisher then moves to a discussion of natural selection. The focus here
is on fitness. This is a crucial element to the unification of Darwin's and
Mendel's theories since it provides a crucial variable that disrupts the equilibrium of axiom 5. It is crucial to be aware that Fisher is examining *groups*
of individuals. His "fundamental theorem of Natural Selection," which
he develops in this section, makes no sense except as applied to groups.
Fisher sets up his formalization (Fisher 1930, 37):[25]

> The two groups of individuals bearing alternative genes, and consequently
> the genes themselves, will necessarily either have equal or unequal rates of
> increase, and the difference between the appropriate values of m will be
> represented by a, similarly the average effect upon m of the gene substitu-
> tion will be represented by a. Since m measures fitness to survive by the
> objective fact of representation in future generations, the quantity $pqaa$ will
> represent the contribution of each factor to the genetic variance in fitness;
> the total genetic variance in fitness being the sum of these contributions,
> which is necessarily positive, or, in the limiting case, zero. Moreover, any
> increase dp in the proportion of one type of gene at the expense of the

---

[23] Although I consider these compelling reasons for treating the H–W equilibrium as an axiom, the
central thesis of this chapter does not depend on it. Mendel's first law could be substituted for the
current axiom 5.

[24] Fisher provides a more detailed mathematical account of the thrust of this section in his enlarged
version of the book published in 1958. Here I stay with the 1930 version, since the 1958 additions
change nothing relative to the case mounted here. To the extent that the additions have any relevance, they enhance the case developed here.

[25] His 1958 version differs in important respects. He provides additional steps to the final equation,
he uses α rather than the smaller a, which avoids confusion of the two a's (larger font a and smaller
font a), and, since $p + q = 1$, he expresses equations in terms solely of p: $\Sigma (a\ dp) = dt\Sigma\Sigma'(2pa\alpha)$
$= W\ dt$.

other will be accompanied by an increase *adp* in the average fitness of the species, where *a* may of course be negative; but the definition of a requires that the ratio *p:q* must be increasing in geometrical progression at a rate measured by *a*, or in mathematical notation that

$$d/dt \log (p/q) = a$$

which may be written

$$(1/p + 1/q)dp = a\, dt,$$

or

$$dp = pqa\, dt.$$

Whence it follows that,

$$A\, dp = pqaadt.$$

And, taking all factors into consideration, the total increase in fitness is

$$\Sigma\, (a\, dp) = \Sigma(pqaa)dt = W\, dt.$$

When *dt* is positive, *W dt* will also be positive. Fisher asserts that, "the rate of increase in fitness due to all changes in gene ratio is exactly equal to the genetic variance of fitness *W* which the population exhibits" (Fisher 1930, 37). He then asserts his fundamental theorem of Natural Selection (*The rate of increase in fitness of any organism at any time is equal to its genetic variance in fitness at that time*), which he compares to the second law of thermodynamics and maintains that it, like that law in physics, holds "the supreme position among the biological sciences" (37). Hence, it would be another axiom.

His discussion of the fundamental theorem is very cryptic and, as a result, Fisher's fundamental theorem has been the subject of much controversy. Many have taken him to be claiming that population fitness will always increase since the theorem seem to be stating that the rate of population increase is equal to the additive genetic variance for fitness. That would entail that population fitness will always increase, which, in fact, is only the case with single locus selection. James Crow (2002) provides a more nuanced and quite defensible interpretation of the fundamental theorem. He understands Fisher's claim to be restricted to the additive genetic component, with no claim about population fitness, in which case *M* (mean fitness of the population) could increase or decrease.

Whatever the fate of the fundamental theorem, the machinery required for unification has been developed and, even without the fundamental theorem, is adequate. Axiom 5 is the starting point for unification:

$$p^2(a_1):2pq(a_1a_2):q^2(a_2).$$

As already noted, this is a cornerstone of a mathematical theory of the inheritance of individual traits (Darwin's axiom 4), and it embodies the maintenance of individual trait differences at the genetic level (Darwin's axiom 3). One thing that will disrupt the equilibrium – and the maintenance of variation – is natural selection, which results from the struggle for existence (Darwin's axiom 2).[26] Fisher's W provides the required mathematical variable for natural selection (the current convention uses w [lower case] for fitness and $\bar{w}$ for mean fitness). Unification only requires that fitness and the H–W equilibrium be mathematically related. This is easily done. If a_1a_1 has a fitness of w_{11}, a_1a_2 has a fitness of w_{12}, and a_2a_2 has a fitness of w_{22}, then, after selection:

$$w_{11}(p^2)a_1a_1:w_{12}(2pq)a_1a_2:w_{22}(q^2)a_2a_2.$$

Normalizing to make $p + q = 1$, yields:

$$(w_{11}(p^2)/\bar{w})a_1a_1:(w_{12}(2pq)/\bar{w})a_1a_2:(w_{22}(q^2)/\bar{w})a_2a_2,$$

where $\bar{w}$ = average fitness. Therefore

$$w_{11}(p^2) + w_{12}(2pq) + w_{22}(q^2).$$

The deductive consequences of this formalization of the unification of Darwin's and Mendel's theories is that evolution is continuous with natural selection acting on small individual variations, and that, in steady state, variation will be preserved; also, even the controversial phenomena of mimicry becomes tractable as a deductive consequence.

That resolved the controversies over continuous versus discontinuous evolution, and the maintenance or decline of variation; something Darwin's informal theory could not do. Only when Darwin's axioms had been formalized was the deduction of continuous evolution and the preservation of variation possible.

REFERENCES

Bateson, William (1902) *Mendel's Principles of Heredity: A Defence.* Cambridge University Press.

---

[26] Today, we know that there are many other drivers of natural selection (fitness) but the struggle for existence is still important.

Bateson, William and Saunders, E. R. (1902) *Reports to the Evolution Committee of the Royal Society.* London: Harrison and Sons.

Crow, James F. (2002) "Perspective: Here's to Fisher, Additive Genetic Variance and the Fundamental Theorem of Natural Selection." *Evolution* 56 (7): 1313–16.

Darwin, Charles (1839) *Journal of Researches into the Geology and Natural History of the Various Countries Visited by H.M.S. "Beagle".* London: George Routledge and Sons (also known as *The Voyage of the "Beagle"*).

(1859) *On the Origin of Species by Means of Natural Selection, or the Preservation of Favoured Races in the Struggle for Life.* London: John Murray.

(1868) *The Variation of Animals and Plants under Domestication,* 2 vols. London: John Murray.

Darwin, Francis (1959) *The Life and Letters of Charles Darwin,* 2 vols. New York: Basic Books.

Fisher, Ronald A. (1930) *The Genetical Theory of Natural Selection.* Oxford University Press.

(1958) *The Genetical Theory of Natural Selection,* 2nd rev. edn. New York: Dover Publications.

Galileo, G. (1623) *Il saggiatori.* Reprinted in English translation in C. D. O'Malley and S. Drake (eds.), *Controversy on the Comets of 1618.* Philadelphia: University of Pennsylvania Press, 1960.

Galton, Francis (1869) *Hereditary Genius: An Inquiry into its Laws and Consequences.* London: Macmillan.

(1872) "On Blood-Relationship." *Proceedings of the Royal Society* 20: 394–402.

(1875) "A Theory of Heredity." *Contemporary Review* 27: 80–95.

(1876) "A Theory of Heredity." *Journal of the Anthropological Institute of Great Britain and Ireland* 5: 329–48.

Gayon, Jean (1998) *Darwinism's Struggle for Survival: Heredity and the Hypothesis of Natural Selection.* Cambridge University Press.

(2008) "Is a New and General Theory of Evolution Emerging? A Philosophical Appraisal of Stephen Jay Gould's Evaluation of Contemporary Evolutionary Theory." In W. Gonzalez (ed.), *Evolutionism: Present Approaches.* La Coruña, Spain: Netbiblo, pp. 77–105.

Gillham, Nicholas W. (2001) *A Life of Sir Francis Galton: From African Exploration to the Birth of Eugenics.* New York: Oxford University Press.

Gould, S. J. (1980) "Is a New and General Theory of Evolution Emerging?" *Paleobiology* 6 (1): 119–30.

Halliburton, Richard (2004) *Introduction to Population Genetics.* Upper Saddle River, NJ: Pearson Prentice-Hall.

Hardy, G. H. (1908) "Mendelian Proportions in a Mixed Population." *Science* n.s. 28: 49–50.

Hartl, Daniel L. and Clark, Andrew G. (2007) *Principles of Population Genetics,* 4th edn. Sunderland, MA: Sinauer Associates.

Herschel, J. F. W. (1831) *Preliminary Discourse on the Study of Natural Philosophy.* London: Longman, Rees, Orem, Brown and Green.

Mendel, G. (1865) "Versuche über Pflanzenhybriden." *Verhandlungen des Naturfor-schenden Vereins in Brünn* 4: 3–47 (all references are to Fisher's reprinting and modification of Bateson's English translation, W. Bateson, *Mendel's Principles of Heredity*. Cambridge University Press, 1909).

Morrison, Margaret (2002) "Modelling Populations: Pearson and Fisher on Mendelism and Biometry." *British Journal for the Philosophy of Science* 53: 39–68.

Pearson, Karl (1904) "Mathematical Contributions to the Theory of Evolution. XII. On a Generalised Theory of Alternative Inheritance, with Special Reference to Mendel's Laws." *Philosophical Transactions of the Royal Society A* 203: 53–86.

(ed.) (1924) *The Life, Letters and Labours of Francis Galton*. Cambridge University Press.

Provine, William B. (1971) *The Origins of Theoretical Population Genetics*. University of Chicago Press.

Punnett, R. C. (1908) "Mendelism in Relation to Disease." *Proceeding of the Royal Society of Medicine (Epidemiology Section)* 1: 135–68.

Ruse, Michael (1973) *The Philosophy of Biology*. London: Hutchinson.

(1975) "Darwin's Debt to Philosophy: An Examination of the Influence of the Philosophical Ideas of John F. Herschel and William Whewell on the Development of Darwin's Theory of Evolution." *Studies in History and Philosophy of Science* 6: 159–81.

(1979) *The Darwinian Revolution: Science Red in Tooth and Claw*. University of Chicago Press.

Stauffer, R. C. (ed.) (1987) *Charles Darwin's Natural Selection: Being the Second Part of His Big Book Written from 1856–1859*. Cambridge University Press.

Weinberg, Wilhelm (1908) "Über den Nachweis der Vererbung beim Menschen." *Jahreshefte des Vereins für vaterländische Naturkunde in Württemberg* 64: 368–82.

Weisheipl, J. A. (1967) "Galileo and His Precursors." In E. McMullin (ed.), *Galileo: Man of Science*. New York: Basic Books.

Whewell, W. (1840) *Philosophy of the Inductive Sciences*. London: Parker.

Yule, G. Udny (1902) "Mendel's Laws and their Probable Relations to Intra-racial Heredity." *New Phytologist* 1: 193–207.

(1903) "Professor Johannsen's Experiments in Heredity." *New Phytologist* 2: 235–42.

CHAPTER 7

Population genetics, economic theory, and eugenics in R. A. Fisher

Jean Gayon

A number of authors have underlined the relationship between Fisher's contributions to population genetics and his eugenic commitments. Some (e.g. Bennett) have minimized the relationship between the two aspects; others have emphasized it (e.g. Norton, Mackenzie). In this chapter, I analyze a special aspect of this relationship: Fisher's use of economic analogies and concepts, both in population biology and in his eugenic thinking. I show that economics offered him a way of articulating his scientific approach to evolution and his eugenic commitments.

Fisher and "economics": overview

Fisher's collected papers (Bennett 1971–74) contain 294 articles. Fisher also published six books, five of which are quite technical (applied mathematics), 59 additional notes listed by his daughter, Joan Fisher Box,[1] and approximately 230 reviews (200 for the *Eugenics Review*). Excluding the five books dealing specifically with mathematical methods, I have browsed through the 353 articles, and one book, *The Genetical Theory of Natural Selection* (*GTNS*), looking for anything that looked like reflections on economic problems, or that made explicit use of economic metaphors. In this first section, I content myself with listing and quickly classifying the texts that will be analyzed in detail in the following sections.

In *GTNS*, Fisher's use of economic concepts falls into two distinct categories:

- *Theoretical*: In chapter 2, devoted to the "Fundamental theorem of natural selection," Fisher compares the growth of a population and the growth of capital invested at compound interest (26–28). This represents

---

I am deeply indebted to Abigail Lustig for her linguistic revision of this chapter, and for her fruitful comments and suggestions.
[1] Fisher Box 1978.

only a few lines, but is crucial for the definition of the Malthusian parameter and the notion of "reproductive value," and, consequently, for the fundamental theorem itself.

- *Practical*: Chapters 8 to 12 are entirely devoted to Fisher's reflection on human evolution and eugenics. Economic arguments pervade the entirety of this reflection. This represents approximately two-fifths of the book.

If we now look at the rest of Fisher's publications, we find 13 articles and at least 2 reviews that make use of economic concepts and offer a substantial "economic" reflection (among the 294 articles and 230 reviews). They correspond to the two categories observed in *GTNS*: theoretical and practical. I say "at least," because occasional use of economic concepts can be found in some other articles or (more probably) in book reviews, but without offering significant developments.

- On the theoretical side, we find an article published in 1927: "The Actuarial Treatment of Official Birth Records,"[2] a major contribution to demography, where Fisher first introduced his "Malthusian parameter."
- On the other side, we find 12 articles both bearing on eugenics *and* using economic arguments (I have not listed all the articles bearing on eugenics). Most of the titles are self-explanatory:
  - "The Eugenic Aspect of the Employment of Married Women" (with C. S. Stock, 1915)
  - Review of L. Darwin, *Eugenics in Relation to Economics and Statistics* (1920)
  - "The Effect of Family Allowances on Population: Some French Data on the Influence of Family Allowances on Fertility" (1927)
  - "The Differential Birth Rate: New Light on Causes from American Figures" (1928)
  - "Review of E. Rathbone, *Ethics and Economics of Family Endowment*" (1928)
  - "Income-Tax" (1928)
  - "Income-Tax Rebates: The Birth-Rate and Our Future Policy" (1928)
  - "The Over-Production of Food" (1929)
  - "The Biological Effects of Family Allowances" (1931)
  - *The Social Selection of Human Fertility* (1932)
  - "Family Allowances in the Contemporary Economic Situation" (1932)

---

[2] Fisher 1927a.

- "Income-Tax and Birth Rate: Family Allowances" (1936)
- "Citizens of the Future. Burden of Falling Birth Rate: The Case for Family Allowances" (1939).

In the texts belonging to the second category, Fisher abundantly used the adjective "economic." All these papers, like chapters 7–12 of *GTNS*, deal with a single "macro-economic"[3] problem: differential birth rates among social classes, and family allowances as a tool for a eugenics policy. Nowhere in Fisher's writings have I been able to find any discussion relating the foundations of evolutionary theory with the foundations of economic theory.

In addition to these data, I would like to add that Fisher published a number of papers devoted to statistical mathematics in several economics journals (such as the *Economic Journal*, *Econometrica*, etc.). This may explain his familiarity with economic problems, but, so far as I can judge, these papers fall outside of the present enquiry: they do not genuinely deal with economic problems.

From now on, I examine whether there is a relationship between the two kinds of texts that I have identified: the theoretical contributions that use the economic metaphor of the growth of capital under compound interest in order to illustrate the demographic problem of the growth of a population, and the texts where Fisher discusses human evolution and eugenics in economic terms.

Malthusian parameter: growth of population and growth of capital

To illustrate the economic flavor of Fisher's treatment of evolution, I will first discuss his concept of fitness, which itself comes at the beginning of the section on the "fundamental theorem of natural selection," probably the most famous passage of *GTNS*. Properly speaking, Fisher does not define fitness; he indicates its relationship with the "Malthusian parameter": "m measures fitness by the objective fact of representation in future generations."[4] In this formula, "m" is the "Malthusian parameter," a term Fisher coined. It designates the relative rate of increase of a population.[5] The word "contribution" has had a striking fate among evolutionists. After Fisher, it became common to characterize fitness as "the contribution a

[3] Macro-economics deals with the structure and behaviors of the entire economy.
[4] Fisher [1930] 1958, 37.
[5] M can be positive or negative; it can be applied to any trait, especially any genetic factor that "contributes" to the overall fitness.

given gene makes to the gene pool of the next generation."[6] Peter Medawar
has crudely expressed the economic connotation involved in the concept
of genetic fitness: "The genetical usage of 'fitness' is an extreme attenuation
of the ordinary usage: it is, in effect, a system of pricing the endowments
of organisms in the currency of offspring, i.e. in terms of net reproductive
performance."[7] Medawar's comment is justified, but Fisher did not explicitly make a parallel with economics in his famous sentence about fitness
in *GTNS*. At most, the verb "contribute" may have a slight economic connotation. In fact, Fisher's explicit economic analogy comes in the context
of his definition of the "Malthusian parameter." I will then concentrate on
this notion rather than on fitness.

The first four sections of chapter 2 of *GTNS* are devoted to a demographic problem. Fisher explains that in order to develop a mathematical
theory of natural selection, one should take into account data about "all
stages in the life-history of an organism."[8] For this purpose, one must be
able to take into account both the chances of survival and of reproduction
at all ages. This is why Fisher introduced the "Malthusian parameter." Let
me recall the main steps of his definition.

Let l_x be the probability of surviving at age x.
Let b_x be the rate of reproduction at age x.

Then, in an infinitesimal element of age dx, the expectation of offspring is
$l_x b_x dx$; over the whole life, the total expectation of offspring is:

$$\int_0^\infty l_x b_x \, dx$$

This quantity gives "the expectation of offspring of the newly-born child"
(Fisher [1930] 1958, 25).

Now, if the age distribution remains unchanged, the relative rate of
increase (or decrease) at all ages will be the same. Fisher calls "m" this
relative rate of increase. At any period, the contribution to the total rate
of increase of persons in the interval dx is $e^{-mx} l_x b_x dx$; e^{-mx} is "the rate at
which births were occurring at the time persons now of age x were being
born."[9] The aggregate for all ages is:

$$\int_0^\infty e^{-mx} l_x b_x \, dx$$

---

[6] Mayr 1982, 588.
[7] Medawar 1960, 108, quoted in Williams 1966, 158.
[8] Fisher [1930] 1958, 22.   [9] Fisher [1930] 1958.

If $\int_0^\infty e^{-mx} l_x b_x dx = 1$, then we have an equation for m, of which there is only one real solution. Fisher explains that he has chosen the symbol m because it designates "the Malthusian parameter of population increase."[10] It is precisely at this stage that Fisher makes an economic analogy:

> In view of the close analogy between the growth of a population supposed to follow the law of geometric increase, and the growth of capital invested at compound interest, it is worth noting that if we regard the birth of a child as the loaning to him a life, and the birth of his offspring as a subsequent repayment of the debt, the method by which m is calculated shows that it is equivalent to answering the question – At what rate of interest are the repayments the just equivalent of the loan? For the unit investment has an expectation of return $l_x b_x dx$ in the time interval dx, and the present value of this repayment, if m is the rate of interest, is $e^{-mx} l_x b_x dx$; consequently the Malthusian parameter of population increase is the rate of interest at which the present value of the births of offspring to be expected is equal to unity at the date of birth of their parent.[11]

The same equations, as well as the economic analogy, had been already presented in Fisher's 1927a, "The Actuarial Treatment of Official Birth Records." I will return later to this important paper.

Fisher's economic analogies did not stop there. After defining the Malthusian parameter, Fisher introduced another parameter, the "reproductive value." I will not analyze here in detail the mathematical complications brought by this concept, which has generated a number of perplexed comments by able mathematicians (both about inconsistencies and about Fisher's intention[12]). What I want to underline is that Fisher also explained his concept of reproductive value through an economic analogy. The "reproductive value" is more than the value of the newly born; it is the present value of the future offspring of a given person at a given age. In Fisher's terms: the "reproductive value" answers the question: "To what extent will persons of this age, on the average, contribute to the ancestry of future generations? The question is one of some interest, since the direct action of Natural Selection must be proportional to this contribution."[13] Fisher added that the reproductive value varies with age, and made this puzzling comment: "[A] positive rate of interest gives higher value to the immediate prospect of progeny of an older woman, compared to the more

---

[10] A few years earlier, Lotka had derived a similar equation; instead of "m" he had used "r" and had defined this parameter as "the natural (or intrinsic) rate of increase population of the population" (see Lotka 1956, 25).

[11] Fisher [1930] 1958, 26.   [12] See Price 1972.   [13] Fisher [1930] 1958, 27.

remote children of a young girl."[14] In a letter to Leonard Darwin of June 1929, Fisher was more explicit: "The reproductive value at different ages must determine the extent to which parental care pays." For instance, in a forest, an old oak may have a reproductive value higher than a younger tree: "It would be a bad bargain for the father oak to benefit his offspring unless he could do so by losing considerably less than the offspring gain."[15] Thus Fisher's demographic approach to natural selection relied upon an explicit economic analogy: making children is interpreted in terms of investment, cost, benefit, and repayment[16]. Before I move to the possible practical stakes (especially eugenical) of Fisher's analogy, I would like to make three comments from a theoretical point of view.

(1) As far as I know, Fisher was not interested in microeconomics;[17] especially, he does not seem ever to have manifested any interest in the concepts and theories of neo-classical economics.[18] Therefore we should *not* attempt to interpret Fisher's analogy in the terms of the modern debates about optimality in evolution and economics. Fisher was interested in macro-economics, especially all questions relating to public policies able to counteract the "degeneracy" of modern populations.

(2) From the point of view of the history of economics, Fisher's analogy between the growth of population and the growth of capital is not trivial. Economists reading Fisher's text immediately recognize a particular idea of the concept, which appeared precisely in the interwar period: capital is a quantity defined, not by actual goods, but by the expectation of future income. According to this notion, the value of a

---

[14] Fisher [1930] 1958, 28.

[15] Bennett 1983, 25, 104. In the same correspondence, Fisher offers another striking example: taking the example of crocodiles, he adds: "I suppose they would co-operate with them not only on terms of mutual advantage, but on terms of joint advantages so long as the loss of either did not exceed the gain of the other. Hence society starts with family" (Bennett 1983, 25, 105).

[16] Note that economic analogies occur only in the first sections devoted to questions of demography. In the following sections of chapter 2 of *GTNS*, Fisher turned to the genetical aspect of natural selection, and asked what happens if the Malthusian parameter is projected upon genetically different individuals in the population. This led him to derive his "fundamental theorem of natural selection." In this part of his argument, Fisher did not use economic analogies. Rather, he presented his theorem as a law occupying "the supreme position among the biological sciences," and he compared it to the second principle of thermodynamics. Just as the law of entropy, the fundamental theorem is about a quantity that always increases. I will not comment here on this aspect.

[17] That is, that part of economics that considers the decisions made at a low level, individuals or firms.

[18] However, I may be wrong, because Fisher was always extremely reluctant to provide the actual sources of his reflection in all domains. In a further paper, I will try to offer more substantial data and conjectures on Fisher and economists.

capital is based, not on an actual quantity of something (of whatever sort) that you own – land, machines, money – but on the returns that you expect if you make an investment on this property. This notion of capital was indeed new in the 1920s. So far, I have been unable to understand whether Fisher was aware of this, or if he had just picked up a popular notion available in current financial practice. Further evidence is needed to properly understand this point.

(3) Finally, from a biological point of view, Fisher's economic analogy comes into play only when he discusses the Malthusian parameter. There is no explicit economic analogy when Fisher discusses natural selection and fitness.[19]

Human evolution, economics, and eugenics

I will now show that Fisher's economic interpretation of the Malthusian parameter was motivated, or at least inspired, not by abstract considerations about natural selection, but by his eugenic way of thinking.

A decisive argument is the 1927 paper where Fisher first elaborated the concepts of "Malthusian parameter" and "reproductive value," and the equations accompanying these notions. This paper was entitled "The Actuarial Treatment of Official Birth Records." The notions developed

---

[19] Note also that the economic analogy comes into play only with respect to fertility, not viability. This is confirmed by the only other example of a similar analogy in the theoretical part of *GTNS* (chapters 1–7), the discussion about sex ratios (chapter 6). There, Fisher explains that the total reproductive value of the males is strictly equal to the total reproductive value of the females, "because each sex must supply half of the ancestry of all future generations of the species." The entire section is written with respect to the "parental expenditure" that would be implied by a sex with a hereditary tendency to produce a bigger proportion of individuals of its own category. "The sex ratio will so adjust itself, under the influence of Natural Selection, that the total parental expenditure incurred in respect of children of each sex, shall be equal; for if this was not so and the total expenditure incurred in producing males, for instance, were less than the total expenditure incurred in producing females, then since the total reproductive value of the males is equal to that of the females, it would follow that those parents, the innate tendencies of which caused them to produce males in excess, would, for the same expenditure, produce a greater amount of reproductive value; and in consequence would be the progenitors of a larger fraction of future generations than would parents having a congenital bias towards the production of females." Therefore, if males are more expensive to produce (for example, because they are more fragile, as is the case in humans), one may expect that natural selection will tend to equate the numbers of males and females that attain the age of reproduction. In passing, Fisher's reflection about the sex ratio makes clear why "reproductive value" (as opposed to the mere "Malthusian parameter") was so important to him. Reproductive value implies taking into account the fertility of progeny; it expands over *more* than two generations (parents and their immediate progeny [Fisher [1930] 1958, 158–60]). One may mention here former expressions of that idea, such as the "progeny test" used by animal and plan breeders since at least Bakewell, or Vilmorin's "genealogical selection" in the mid nineteenth century. I am grateful to Abigail Lustig for this observation.

there resemble tremendously the first pages of *GTNS*, chapter 2. But there is a big difference between the 1927 paper and *GTNS*: the 1927 paper does not contain the slightest allusion to evolution and natural selection. Therefore, the 1927 paper offers us an exceptional tool for understanding what was originally at stake in Fisher's concepts of Malthusian parameter and reproductive value. These notions were not first intended to elaborate the mathematical theory of natural selection.

The aim of the 1927 paper was to suggest better indices for future censuses. Better censuses, Fisher said, should help obtaining a better image of "the reproductive history of the immediate past."[20] As I remarked above, in this article, Fisher did not allude to the theory of natural selection. But all the elements constituting the first three sections of *GTNS*, chapter 2, are there: life tables, tables of reproduction, Malthusian parameter, analogy with the rate of compound interest, and "reproductive value of a man at age x." What, then, was at stake? What mattered for Fisher was the comparative reproductive value of human groups, with respect to "'occupations' or to any typical trait that affect their vitality." The following quotation shows well what Fisher had in mind:

> [T]the years subsequent to the cessation of active work are … unimportant biologically … if, however, we are concerned with the mortality of any class in relation to its natality,[21] or activity in reproduction, it is obvious that the deaths of children should be included in the life table according to the occupations of their fathers, for their real life table is that of a diminution of the net fertility of this class.[22]

And, a few lines later:

> The expectation of offspring of a living newly born child subjected to the contemporary conditions of mortality and natality provides the simplest … measure of the natural increase or decrease in the class considered.[23]

Thus, Fisher's claim is that human groups, especially groups defined by "occupations" (or social classes), are unequal with respect to their "reproductive value." Fisher's concern was about the real capacity of different social classes to reproduce enough, and early enough, to contribute to the

---

[20] Fisher 1927a, 108.

[21] Natality is a synonym for birth rate: it is measured by the ratio of the number of births to the size of the population.

[22] Fisher 1927a, 105.

[23] Fisher 1927a, 105. In the same mood: "For most occupations in England and Scotland the writer believes that … we should find the population decreasing, but it is characteristic of our present ignorance that even a point of such prime importance cannot be determined at once from official data" (Fisher 1927a, 104).

general increase of the population. It is precisely at that point that Fisher refers to Malthus: "The scheme of ideas for the exact treatment of this factor is included in Malthus' analogy of population increase with compound interest. If capital is to be repaid at a premium the date of repayment is of importance in the calculation of the rate realized."[24] This being said, the 1927 paper is a technical one, where the influence of eugenic motivations can only be guessed at. To identify this eugenic motivation, one has to look at the popular papers where Fisher developed his ideas about the decline of civilization, the social selection of human fertility, and family allowances.

I will now examine Fisher's major eugenic theses with special respect to their economic aspects. Fisher's eugenic doctrine comes down to a few leitmotivs, almost all of which have a strong economic flavor.

The most distinctive feature of Fisher's eugenic ideas is his typical habit of combining a Darwinian approach to human evolution with economic considerations.[25] Here are the main leitmotivs of this evolutionary-and-economic view of eugenics.

(1) Like Darwin, Fisher was fascinated by the question of the origins of cooperative behavior and altruism. Like Darwin, too, he emphasized the role of intergroup selection. But in contrast to Darwin, who wanted to solve the problem of the origins of the moral sense and of social virtues, Fisher was interested in the economic aspects of cooperation. He claimed that "the economic system" was one of the most impressive triumphs of early human organization. By "economic system" he meant "the free interchange of goods and services between different individuals whenever such interchange appears to both parties to be advantageous."[26] In such a system, "each individual is induced, by enlightened self-interest, to exert himself actively in whatever ways may be serviceable to others."[27] Those who succeed receive the highest rewards; those who fail to perform socially advantageous actions have poor access to the wealth of the community. This view of the economic system – Fisher says – is in full agreement with both "the theory of rationalist economists, and ... the practice of various ancient civilizations."[28]

[24] Fisher 1927a, 105.
[25] A number of historians have emphasized one aspect that has nothing to do with economics, the postulate of the inheritance of most human qualities, physical, mental, and moral. Hereditarianism is certainly a key component of Fisher's eugenic conceptions. As shown by historians of science such as Bernard Norton (1975, 1978) and Donald Mackenzie (1981), hereditarianism played an important role in Fisher's involvement in the construction of quantitative genetics. But hereditarianism is not distinctive of Fisher's eugenic conceptions. All eugenicists believed that human behavior and human mental and moral abilities were hereditary at a high degree.
[26] Fisher [1930] 1958, 202–3. [27] Fisher [1930] 1958, 202. [28] Fisher [1930] 1958, 202.

Fisher admitted that the economic system was not the exclusive basis of sociality in man: social instincts and moral sense also played a role. But for him, the economic system was the most distinctive aspect of the evolution of cooperation in the human species.

What interested Fisher were the evolutionary consequences of "the economic system." Fisher insisted that the economic mode of cooperation creates problems that do not occur in animal societies based upon mere instinct. In a community of social insects, there is no "intra-communal selection": the evolution of communities relies on "inter-communal selection," or, as Darwin would have said, "community selection." In human societies, Fisher said, it is the opposite: differences of wealth interfere with differential rates of death and reproduction, and therefore entail a constant and rapid evolutionary change within populations.

(2) So far, eugenics is not yet visible in Fisher's speculation. The properly eugenic element results from a comparison between primeval societies. In primeval societies, Fisher says, economic success converges with reproductive success: the most socially able individuals are also those who reproduce best. But in "civilized societies," socioeconomic aptitudes and fertility are decoupled. The higher social classes reproduce less. Fisher emphasized that Malthus's views about overpopulation had been disconfirmed by the evolution of modern societies.[29] For Fisher, Malthus's argument had been contradicted on both sides: on the one hand, since the agricultural revolution, the production of food has increased, not in an arithmetical, but in a geometrical ratio; on the other hand, the generalization of wages has induced people to restrain their fertility. Modern societies give extra value to the absence of children. Having children is costly, but wages are the same for people who do or do not have children: "In giving, through the wage system, a bonus to childlessness, we are not procuring the community for nothing. We are buying childlessness in the same market as we are buying the competence, and other qualities, needed to produce wealth."[30] Here we begin to see the rationale behind the analogy between growth of a capital and the growth of a population.

(3) This led Fisher to develop a theory of social promotion, a popular subject among social scientists in the first half of the twentieth century, especially those who opposed socialism and communism. Fisher's idea was that "those whose social promotion has been most striking have on the average fewer children than those whose social promotion has been less."[31] Here

[29] Fisher 1929.   [30] Fisher 1929, 57.   [31] Fisher 1928b, 184.

is Fisher's subtle argument: social aptitudes, which are decisive for obtaining wealth, are unequally shared among individuals; social aptitudes have an important hereditary component; therefore, social aptitudes will tend to concentrate in social groups that are defined both in terms of hereditary endowment and economic status. But social classes are not static entities; in modern societies, they are permanently renewed through a process of social promotion; moreover, the most socially able are also the least fertile; therefore, in each generation the least fertile individuals are given a social and economic advantage over other individuals. Therefore, modern societies constantly recruit their elite from the least fertile individuals among those of their original class, a process that impedes progressive evolution.[32]

Fisher concludes:

> The system of social promotion shows itself as a machine acting with the automatic certainty of natural law, uniting the highest form of ability with relative sterility or defectiveness of the reproductive instincts, at one end of the social scale; and, at the other, the lowest grades of utility with all genetic determinants of high fertility.[33]

Fisher thought that this was genuinely an evolutionary process, and, more importantly, an extremely rapid one, able to weed out half of the "ability" at each generation.

(4) The fourth leitmotiv is the remedy for this degenerative evolution. For approximately 12 years, from 1927 to 1939, Fisher publicly advocated family allowances as the appropriate means to oppose the decline of fertility in the higher classes. This was the period when family allowances were being hotly debated in the United Kingdom, after the French had adopted them as a means of combating their decline in birthrate.

Fisher's attitude towards family allowances was consonant with the official position of the Eugenic Society, which he liked to recall: "The aim of such a system [a scientifically designed of family allowances] should be to equalize the standard of living between parents and non-parents doing equivalent work, within all grades affected, in such a way that the amounts recovered per child by each class of earner shall be proportional to the earnings" (*Outline of a Practical Eugenic Policy*, quoted in Fisher 1932b). In all his reflections on family allowances, Fisher

---

[32] "When we consider ... that in *all* ranks of society, *all* cases of net infertility, whether physiological merely or acting through voluntary choice, must in each generation be given a social advantage over the opposite qualities, borne by persons of equal natural ability, it is evident that the more affluent classes of society must have become differentiated from the poorer, as much in their low innate fertility as in their high natural ability" (Fisher 1932a, 28–29).

[33] Fisher 1932a, 29.

insisted on the necessary dual perspective – economic and evolutionary – within which family allowances are to be considered: "Family allowances should have come to be advocated for a primarily biological object, while the economic effects should be regarded as accidental concomitants ... Economic prosperity itself depends on how many, and still more on 'of what kind' we are."[34] This is why Fisher and the Eugenics Society advocated a "non-flat" policy of family allowances. Family allowances should guarantee "equal standard of living for equal work," a principle meaning that the amount of family allowances should be modulated according to salary, in order to preserve the potential to reproduce. This is why Fisher expressed his opposition to the French system of family allowances a number of times.

Conclusion

I would like to suggest a link between the two aspects of Fisher's "economic way of thinking," theoretical and practical. The puzzling analogy that Fisher made between the growth of a population and the growth of capital in *GTNS* had indeed something to do with his eugenic commitments. I think it is not by chance that Fisher construed the birth of a child as "the loaning to him a life," and the birth of the offspring of this child as "a subsequent repayment of the debt."[35] For Fisher, reproduction was inextricably both an evolutionary problem and an economic problem. It should be emphasized, however, that the link between Fisher's treatment and his eugenic commitments makes sense only at the level of the motivations and the ideology that intervened in the genesis of his thinking. The formal analogy between the growth of a population and the growth of capital under compound interests would still hold even if Fisher had not been a eugenicist. What I have found fascinating in the story I have tried to reconstruct is the role of ideology at the heart of Fisher's most theoretical contributions in demography and evolution. We face here something resembling the role of alchemy and theology in Newton's discovery of the principle of universal attraction. In Fisher, it is not theology, but a major political ideology that penetrated at the heart of his work on evolution. I am glad to dedicate this chapter to Michael Ruse, whose interest and respect for intellectual history has always been on a par with his interest for evolution and philosophy of biology.

[34] Fisher 1931, 21.   [35] Fisher [1930] 1958, 27.

REFERENCES

Bennett, J. H. (ed.) (1971–74) *Collected Papers of R. A. Fisher*, 5 vols. University of Adelaide.

(1983) *Natural Selection, Heredity and Eugenics, including Correspondence of R. A. Fisher with Leonard Darwin and Others*. Oxford: Clarendon Press.

Box, J. Fisher (1978) *R. A. Fisher: The Life of a Scientist*. New York: Wiley.

Fisher, R. A. (1920) "Review of L. Darwin, *Eugenics in Relation to Economics and Statistics*." *Eugenics Review* 11: 60–61.

(1927a) "The Actuarial Treatment of Official Birth Records." *Eugenics Review* 19: 103–8.

(1927b) "The Effect of Family Allowances on Population: Some French Data on the Influence of Family Allowances on Fertility." Family Endowment Society pamphlet, pp. 7–11.

(1928a) "Income-Tax." *Eugenics Review* 19: 231–33.

(1928b) "The Differential Birth Rate: New Light on Causes from American Figures." *Eugenics Review* 20: 183–84.

(1928c) "Review of E. Rathbone, *Ethics and Economics of Family Endowment*." *Eugenics Review* 20: 72.

(1928d) "Income-Tax Rebates: The Birth-Rate and Our Future Policy." *Eugenics Review* 20: 79–81.

(1929) "The Over-Production of Food." *Realist* 1: 45–60.

(1930) *The Genetical Theory of Natural Selection*. Oxford: Clarendon Press, new edition New York: Dover, 1958.

(1931) "The Biological Effects of Family Allowances." *Family Endowment Chronicle* 1: 21–25.

(1932a) *The Social Selection of Human Fertility*. The Herbert Spencer Lecture. Oxford: Clarendon Press.

(1932b) "Family Allowances in the Contemporary Economic Situation." *Eugenics Review* 24: 87–95.

(1936) "Income-Tax and Birth Rate: Family Allowances." *The Times*, April 30.

(1939) "Citizens of the Future. Burden of Falling Birth Rate: The Case for Family Allowances." *The Times*, April 10.

Fisher, R. A. and C. S. Stock (1915) "The Eugenic Aspect of the Employment of Married Women." *Eugenics Review* 6: 313–15.

Gayon, J. (1998) *Darwinism's Struggle for Survival: Heredity and the Hypothesis of Natural Selection*. Cambridge University Press.

Lotka, A. J. (1918) "The Relation between Birth Rate and Death Rate in a Normal Population." *Quarterly Publications of the American Statistical Association* 16: 121.

(1956) *Elements of Mathematical Biology*, New York: Dover (first published 1924).

Lotka, A. J. and I. Dublin (1925) "On the True Rate of Natural Increase of a Population." *Journal of the American Statistical Association* 20: 305–39.

MacKenzie, D. A. (1981) *Statistics in Britain 1865–1930. The Social Construction of Scientific Knowledge*. Edinburgh University Press.

Mayr, Ernst (1982) *The Growth of Biological Thought. Diversity, Evolution, and Inheritance*. Cambridge, MA: Harvard University Press.

Medawar, P. B. (1960) *The Future of Man*. New York: Basic Books.

Norton, B. J. (1975) "Metaphysics and Population Genetics: Karl Pearson and the Background to Fisher's Multi-Factorial Theory of Inheritance." *Annals of Science*, 32: 537–53.

(1978) "Karl Pearson and the Galtonian Tradition: Studies in the Rise of Quantitative Social Biology." Ph.D. thesis, University of London.

Price, G. R. (1972) "Fisher's 'Fundamental Theorem' Made Clear." *Annals of Human Genetics* 36: 129–40.

Williams, G. C. (1966) *Adaptation and Natural Selection: A Critique of Some Current Evolutionary Thought*. Princeton University Press.

CHAPTER 8

Exploring development and evolution on the tangled bank

Jane Maienschein and Manfred Laubichler

When Michael Ruse was young, a linear model of knowledge generation prevailed. Scientists (and philosophers) make discoveries, and generate knowledge, on this view, then society uses (or doesn't use) that knowledge. According to this interpretation, using scientific knowledge usually makes society – and the world – better, though occasionally unfortunate applications make the world worse. In each case, the arrow of influence was thought to point from science to society.

Then after the 1970s, those emphasizing social contexts insisted on the importance of arrows pointing the other way as well, or even primarily. As an example, E. O. Wilson's presentation of sociobiology was seen as his having chosen a view about nature that could have been otherwise. Critics, starting with Wilson's colleagues and former students at Harvard, attacked Wilson, eventually dumping water on him at a meeting to demonstrate their objections. Society, on their view, could and should choose another biological view of human nature. This critique became part of a movement that insisted on understanding science as something that is socially constructed.

Of course, both these simplistic ideas of linear influence are limited and problematic. Instead, a great deal of lateral transfer of ideas also occurs, and what we have is much more nearly a tangled bank of influences at the intersection of science and society. Similarly with philosophical understanding of biology: when Michael Ruse began his career, the picture was much simpler with emphasis on reduction and levels of selection. David Hull's short 1974 textbook *Philosophy of the Biological Sciences* offered only two pages related to development, in particular causal laws of development largely based on genetics and focused on issues of reductionism.[1] In his own 1973 textbook *The Philosophy of Biology*, Michael Ruse acknowledged that he had emphasized evolutionary biology rather than covering

---

[1] Hull 1974.

every biological topic. He noted that "many areas of biology, for example, embryology, have been practically or entirely ignored. My silence about them should not be taken to indicate that I think them of no philosophical significance – I feel sure that, in fact, such areas harbour important problems awaiting discussion."[2] In fact, with that work and subsequent books, Michael Ruse has gone far to help reveal that the bank of biological and of philosophical issues is indeed tangled and that it is worth marveling at the resulting "grandeur" that comes from the complexity and diversity.

The field called evo-devo gives us a nice example of just this sort of tangle of fields and issues, as well as a great deal of lateral transfer. In a blog entry on July 5, 2005, Michael Ruse asked "What are the big issues today?" He noted:

> I pride myself on having a pretty good nose for a problem, and if I were going in the direction of straight philosophy of biology – as opposed to something that was going to bring in history, ever a fondness of mine – I would without hesitation go for evolutionary development, "EvoDevo." I think some of the most incredible discoveries of recent years have come from this area – the amazing homologies between humans and fruitflies for starters.[3]

He went on to suggest that surely evo-devo will have something intriguing to say about human evolution, perhaps focusing on cognition, and therefore that evo-devo can impact our social understanding of ourselves.

In addition, the field that has come to be called evo-devo is also shaped by the society of biologists involved. In this chapter, we look briefly at the backgrounds and origins of the field, at what is at issue today, and at some trends for the future in biology and for philosophy of science amidst the entangled studies of development and evolution.

Background

Darwin provided the first connections of embryology and evolution when he pointed to embryological facts and asserted that embryos provide the strongest support for his ideas of evolution, "second in importance to none in natural history." In his *Autobiography,* he wrote that "Hardly any point gave me so much satisfaction when I was at work on the *Origin,* as the explanation of the wide difference in many classes between the embryo

[2] Ruse 1973, 218.   [3] Ruse 2005.

and the adult animal, and of the close resemblance of the embryos of the same class. No notice of this point was taken, as far as I remember, in the early reviews."[4]

Darwin acknowledged that Ernst Haeckel had done far more with the connections than Darwin had and that it was therefore perhaps understandable that reviewers had not taken much notice of his own contributions. Enough has been written about Haeckel and ideas of recapitulation that we need not go over that ground again but can move on.

In 1896, Edmund Beecher Wilson wrote in what became his classic study of *The Cell in Development and Inheritance*, that "The cell-theory must therefore be placed beside the evolution-theory as one of the foundation stones of modern biology." Yet he recognized that the two were very different fields of study, with different kinds of research and questions. One looked at individual cells from individual organisms, used the microscopic, and spent time inside in the lab, while the other started from natural historical study of behaviors and groups. Both have problems explaining how complex phenomena arise – but to Wilson, at the end of the nineteenth century, it seemed clear that inherited germ-plasm provides a basis for all biological phenomena. Wilson called for careful study of the cell, and also urged that finding intersections of the cell and evolutionary theories, with heredity and development at the core of both, was essential for real progress in the biological sciences.[5]

Some biologists have pointed to work around the end of the nineteenth century as the point of divergence of evolution and development. For example, Rudolf (Rudy) Raff and Thomas Kaufman noted in 1983 that though it seemed obvious that research in evolutionary biology should include a major study of developmental processes, this was not the case. As they put it, "Embryological development, which was so vital a part of evolutionary theory in the late nineteenth century, has been considered largely irrelevant in the twentieth." Despite efforts by Hans Driesch and Thomas Hunt Morgan, which they cite as the two major efforts to join the two different fields, they felt that a truly modern synthesis had failed because of lack of enough understanding of developmental genetics. Yet in 1983, they felt that "the time has come to take the final step in the modern synthesis: To fuse embryology with genetics *and* evolution."[6]

---

[4] Darwin 1859, 450; Barlow 1958, 125. Ernst Mayr especially liked to point to Darwin's use of embryology, for example in Mayr 1982, 469–70.
[5] Wilson 1896.   [6] Raff and Kaufman 1983, 1, 24.

Their perspective is valuable because Raff has become one of the leaders in the established field of evo-devo studies. That he saw the reconvergence of what had been divergent lines of research up to the 1980s or so is reinforced in another volume that he edited with Elizabeth Raff and that resulted from a symposium at the Marine Biological Laboratory – where much of the early work had been carried out. Here, in 1987, the Raffs pointed to two lines of research extending from Haeckel's emphasis on embryology as a way to get information about evolution and phylogenetic relationships on the one hand, and experimental embryology as a way to understand development of individual organisms on the other. For the first two-thirds of the twentieth century, it was very difficult to bring the two lines together, in large part because developmental biologists and evolutionary biologists use different methods, ask different questions, and "practitioners of one discipline are generally unaware of the paradigms and body of knowledge considered central to the second, and thus may have more than a little difficulty posing questions that can be considered nontrivial by members of the second discipline." Philosophical differences – by which they mean differences in epistemologies – ensured that little communication occurred at the intersections.[7]

These founders provide one perspective on what kept the fields apart, but a quick look at some intervening contributors suggests that the story is more complicated and the fields more tangled than that. In a largely ignored paper presented at the bicentennial of the University of Pennsylvania and published in a volume on *Cytology, Genetics, and Evolution* that appeared in 1941, cytologist Clarence E. McClung pointed out that development involves a series of interactions of the organism with its changing environment, which allows multiple possible results that were not fixed by heredity. As McClung put it:

> Perhaps the problem might be stated specifically in this way: Since all organisms exhibit a common series of functions, and since functions are performed under the control of a recognizable series of agents within the chromosomes, there must exist a nuclear mechanism common to all organic types. Logically this would follow, and observation tells us that, at least in its major features, such a situation does exist, for cellular structure and behavior are essentially the same wherever found. Moreover, in the earlier and relatively simpler processes of development, strong likenesses prevail through all forms. But beyond this, what are we to expect? Does each advance in complexity, each new structural element mean additional gene

---

[7] Raff and Raff 1987.

controls or are they due to what might be called the better education of members of an existing system through new experiences?[8]

McClung called for more detailed study of the germ-plasm, and especially the interactions of genetics and cytology to get at the relations to development and evolution. Like Raff, Raff, and Kaufman, McClung recognized the need for more information. But his work reminds us not to assume that people were not trying to bring the fields together. The work also shows that some cytologists and embryologists were beginning by the 1940s to see that developmental processes are often conserved across different species, which suggests a strong hereditary and evolutionary impact on development.

Julian Huxley, in articulating "The Modern Synthesis," saw this as well. It has become popular for historians and biologists to assert that the synthesis ignored development, and many developmental biologists have called for correcting that slight. Yet Huxley did note in a section entitled "The Consequences of Differential Development" that different "rate-genes" control the rate and timing of developmental processes. Therefore, development is a key part of playing out of evolution.[9]

Stephen Jay Gould, in his 1977 *Ontogeny and Phylogeny*, said essentially the same thing, only at much greater length. He noted that his book was a personal favorite of his, on which he had expended a great deal of effort. There he pointed to the intersection of ontogeny and phylogeny as focused on heterochrony – that is, the timing and rate of appearance of characteristics during development, which reflects evolutionary history.[10]

The 1970s and 1980s did begin to reveal a great deal more about the connections. As John Gerhart and Mark Kirschner noted in 1997, a "flood" of new experimental approaches and new results had "convinced everybody" that the underlying developmental mechanisms are conserved.[11] Instead of trying to find what is the same in development of different organisms, the question had shifted to how difference arises.

Scott Gilbert modified his *Developmental Biology* textbook for the 6th edition in 2001, adding chapters on plants, medical implications, environmental regulation, and finally chapter 23 on "developmental mechanisms of evolutionary change."[12] Brian K. Hall and Wendy M. Olson's edited

---

[8] McClung 1941, 64.
[9] Huxley 1943 (dedicated to Morgan, "many-sided leader in biology's advance," 525–43 is a section on "Consequential Evolution: The Consequences of Differential Development").
[10] Gould 1977, 2.   [11] Gerhart and Kirschner 1997, ix.   [12] Gilbert 2006.

collection of *Keywords and Concepts in Evolutionary Developmental Biology* in 2006 further demonstrated that studies at the intersection of evolution and development had come of age.[13] The next steps were to establish the research as a field, to celebrate convergence, and then to get down to the hard work of figuring out what the tangle of ideas and fields really involves and what progress is really being made.

Establishment of evo-devo

The field gradually emerged as multiple different, and somewhat divergent, forces came together. Various different accounts – from participants as well as observers – offer different creation myths, but it is clear that a number of researchers coalesced around a number of related ideas at about the same time.[14]

Here we focus on just one (relatively recent) story that is well documented, and furthermore both of us were there to witness the event. This official stage in the emergence of evo-devo occurred at the annual meeting of the Society for Integrative and Comparative Biology (formerly the American Society of Zoologists) in 2000. SICB had just added a Division of Evolutionary Developmental Biology with Rudolf Raff as the inaugural chair. Richard Burian explained in his Introduction to the published symposium volume that the impetus grew out of the previous SICB meeting in Denver. At a breakfast, he recalled, Burian, Scott Gilbert, Paul Mabee, and Billie Swalla mapped out a symposium to explore the historical background of evo-devo. In fact, they ended up with three symposia, one historical to look at the past, present, and future of the new field, and the others to explore hox genes and "using phylogenies to test hypotheses about vertebrate evolution." They clearly placed their emphasis on evolutionary aspects of developmental biology, and hence the name evo-devo.

Yet differences in underlying assumptions already appeared in the historical symposium entitled "Evolutionary Developmental Biology: Paradigms, Problems, and Prospects." In his essay introducing the special symposium, Burian pointed to biological research at the end of the nineteenth century, when research flourished on cell division, embryology, evolution of species, fertilization, heredity, and phylogeny. "These problems," he noted, "were generally held to be intimately interconnected, so much so that many biologists thought of them as inseparable, forming

---

[13] Hall and Olson 2006.   [14] See Laubichler and Maienschein 2007, 2013.

a single nexus that covered what was eventually separated into cytology, embryology (later transformed into developmental biology), evolutionary biology, genetics, and reproductive biology." He went on: "From a current perspective it is very difficult to understand the ways in which all of these problems were tangled together at the end of the nineteenth century."[15]

Yet in the intervening century, all these areas of study had become specialized fields, each with its own underlying assumptions, different methodologies, and different ideas about the important questions. Studies of evolution of species (seen as relating to populations) and about embryonic development (relating to individual organisms and its parts) grew apart. Departmental organization reinforced the intellectual and methodological differences, so that in many institutions these two areas had little contact by the end of the twentieth century. Philosophy of biology followed the trend, so that those arguing about how to understand evolutionary levels of selection, for example, rarely thought of looking at developmental biology.

The 1990s started to change that trend, as Burian notes. New methods in biology, molecular and genetic techniques, and the discoveries associated with them, widely began to transform studies of both development and evolution. Burian pointed to tools from phylogenetic systematics, the discovery of the conserved nature of transcription factors, the possibility of targeted gene knockouts, and the allure of hox genes as examples. Programs of philosophy of science meetings from the time show that a very few philosophers also began to look at development and ask epistemological questions about how we can best study individual and cellular development, and how that knowledge relates to study of evolution of populations. At the 2000 SICB symposium, papers on modularity and homology focused attention on the potential for intersections of fields.

The symposium ended with two keynote talks; one by Rudy Raff as Chair of the new division, the other by Yale biologist Günter Wagner. While Raff's talk was an affirmation of his previously published views and was ultimately not published in the symposium papers, Wagner and his co-authors (full disclosure – ML was one of them) pointed to underlying differences of view that challenge the claims of evo-devo as the best way to bring together the two fields.[16] In contrast to most papers, which focused on how evolutionary information and theories can be valuable for informing understanding of development, Wagner's team emphasized the

---

[15] Burian 2000.   [16] Wagner *et al.* 2000.

use of developmental mechanisms for understanding evolution. The two perspectives, of evolutionary developmental biology (evo-devo) and of developmental evolution (devo-evo) respectively, ultimately go together, yes, but in more complex ways that recognize the tangle of issues and approaches.[17]

In the initial phase of institutional establishment of evo-devo and devo-evo the epistemological differences underlying these two perspectives also reflected different trajectories in training as well as different historical traditions. As a first approximation we can note that the majority of the proponents of evo-devo were initially trained in developmental biology and developmental genetics and were eager to adopt the insights derived from evolutionary and phylogenetic comparisons – such as the conservation of the hox genes, the focus on the second symposium – to the understanding of development and the evolution of developmental systems.

On the other hand many of the proponents of developmental evolution were trained either in evolutionary theory or evolutionary morphology, or they represented a unique combination of expertise such as developmental and molecular biology or developmental biology and paleontology. They had in common a focus on mechanistic explanations of phenotypic evolution and an interest in the origin of evolutionary innovations and of phenotypic variation more generally. A good number of those were also exposed to or had already contributed to a growing number of challenges to the orthodoxy of the Modern Synthesis of the 1940s.[18]

The boundaries between these two camps were, of course, fluid and not at all rigid, but they nevertheless represented two different epistemologies and sets of questions. These differences in orientation were also visible in the two major journals that came to represent the new field (and which were soon adopted as journals supported by the SICB Division.) Both of them were launched in 1999. *Evolution and Development* was established as a new journal under the editorship of Rudolf Raff, whereas *Molecular and Developmental Evolution* with Günter Wagner as editor-in-chief was first set up as a free-standing section of the *Journal of Experimental Zoology*, which had a long tradition in publishing articles at the intersection between development and evolution. It became an independent publication as the *Journal of Experimental Zoology, Part B: Molecular and Developmental Evolution* in 2003.

[17] Wagner *et al.* 2000.
[18] Laubichler 2009 and Laubichler and Maienschein 2013.

The two "camps" thus each have their own publication and even though there are areas of overlap, a comparison of the papers published in each journal does reveal different centers of gravity that map onto the underlying conceptual differences. The scientific questions investigated within both contexts have over the last decade become more popular and furthermore substantial progress has been made. Therefore an increasing number of publications devoted to both evo-devo and developmental evolution have more recently been published in leading generalist journals, such as *Science, Nature, Cell, PNAS*, etc., thus further blurring the boundaries between these viewpoints.

At the time of setting up the new SICB Division, Wagner was a candidate running for office. In his candidate's statement, he noted the exciting prospects for the new research directions. Yet he also argued that:

> The division is the first formal union of researchers in the field and will have to play a pivotal role in the process of defining the identity of the discipline. This process can only succeed if the division is able to attract the knowledge and talent of developmental biologists, evolutionary geneticists, morphologists, systematists and paleontologists. I strongly believe that each group makes an essential contribution to the developmental evolutionary synthesis. To make the integration of these contributions possible, the division should be perceived as a place where researchers from different backgrounds can communicate without facing disciplinary chauvinism. SICB has a strong tradition in this regard.

In running for office for the next year, he said that he planned to "draw on its accumulated organizational wisdom to reach this goal."[19]

Wagner already recognized some of the social forces within biology that were likely to affect the efforts to form a new field. He envisioned the goal as a "synthesis" while others stressed the value of one field for the other. Much sociological literature has explored the conservative resistance to new ideas, especially in academia. When much hinges on one's own field retaining a place of stature, participants resist letting go of their power. Thus, as we have explained, friendly sparring occurred around the question whether the field is really "evolutionary developmental biology" or should more properly be "developmental evolutionary biology." The two are not the same, obviously, and this was not just a political attempt at priority. Rather, two different conceptions have coexisted from the beginning. Yet, again, it is not just a question of which of the two more

---

[19] See Günter Wagner's statement on the SICB website: www.sicb.org/newsletters/nl04–2000/dedb. php3.

influenced the other; it is not a question of which of two directions the arrows point in. Rather, the arrows are sometimes bent, intertwined, and definitely entangled.

This is especially true if one also takes the more recent developments into account. These include on the one hand a call for the explicit inclusion of ecological factors into the theoretical structure of evo-devo ("eco-evo-devo") often paired with a (mostly) not very precise emphasis on "epigenetics,"[20] while on the other hand we have developments that take advantage of most up-to-date methods of synthetic and molecular biology in order to both experimentally and computationally reconstruct evolutionary transitions – the idea of synthetic experimental evolution (SEE).[21] The landscape of what used to be called evo-devo has thus become its own complex ecosystem and as such an object of considerable attraction for innovative work in the philosophy of biology. And indeed, philosophers and historians have been exploring this territory and have produced a considerable amount of scholarship, often encouraged by Michael Ruse as only he can.[22]

In the remaining part of this chapter we will not dwell on what has been accomplished, rather we will follow Michael Ruse's advice and point towards some of the emerging questions at the intersection of developmental evolution, systems biology, synthetic biology, and computer science, as it is precisely at those "trading zones" between ideas, technologies, and epistemologies that we can best observe the tangled bank of mutual interactions between biology and society.

The future of developmental evolution and its HPS challenges

Arguably, evo-devo and developmental evolution (the version of the problem we have been focusing on for more than a decade) have been among the most exciting areas of the life sciences, especially for a die-hard evolutionist like Michael Ruse. And indeed, there has been much talk about "completing the synthesis" and finally integrating development into evolutionary theory in evo-devo circles. For a historically inclined philosopher, such claims are the equivalent of a five-course meal, if not an "all-you-can-eat" buffet. And despite all the ink spilled by philosophers debunking the myth that evolution is necessarily progressive, the claims of "completing"

---

[20] Gilbert and Epel 2009.   [21] Erwin and Davidson 2009.
[22] See for instance Laubichler and Maienschein 2007 and Laubichler and Maienschein 2009.

and "finally integrating" sound quite teleological. So there is much to question here about the development of scientific theories, what it means for a theory to be complete, or what integration actually stands for.

But these kinds of claims have mostly been associated with evo-devo. Here we will focus on developmental evolution, its future, and the challenges it raises for the historian and philosopher. So let's start at the beginning: more precisely, what are the major substantive differences between evo-devo and developmental evolution?[23]

We have seen that these were on display during the inaugural symposium of the SICB Division. Shortly thereafter Brian Hall wrote a guest editorial in *Evolution and Development* asking the question "Evo-devo or Devo-evo – does it matter?"[24] For Hall it clearly did. He pointed to some of the differences in emphasis between these two perspectives but also suggested that, if one takes all these claims to their logical conclusion then evo-devo presents a synthesis (completing the Modern Synthesis argument) while devo-evo, once fully developed, would represent a revolution. In 2000 much of what developmental evolution would eventually become was not clearly visible – and to be sure, there were, then as now, some differences in Brian Hall's view of devo-evo and in our (Wagner and Laubichler's) conception of developmental evolution. But in many ways Hall's argument that there is indeed a difference and that it matters foreshadowed much of what would happen in the next decade.

Let us now turn to what we see as the main differences between evo-devo and developmental evolution and focus primarily on historical and epistemological issues. On the level of experimental data we observe a convergence between the two approaches; therefore the remaining differences are largely interpretative and conceptual.

Evo-devo, as a synthesis that purportedly completes the efforts of the 1930s and 1940s by finally incorporating development into the theoretical structure of evolutionary theory, does accept the core theoretical assumptions of neo-Darwinian evolutionary biology, namely that all explanations of evolutionary change are ultimately provided by the (adaptive) dynamics within populations. Natural selection and random genetic drift are the processes that govern evolution. Phenotypic change, in this view, is predicated on the underlying dynamics of alleles. And while earlier models assumed that the map between genotypes and phenotypes is simple (linear or additive), new empirical evidence and theoretical considerations

[23] Laubichler 2009.   [24] Hall 2000.

soon challenged this convenient view. Understanding the more complex relationships between genotypes and phenotypes, such as observations on constraints on phenotypic variation and the apparent directionality of evolutionary transformations, then required considerations of development. Within evo-devo, the logical place of development within evolutionary theory was in explaining the details of the genotype–phenotype map without changing the explanatory structure of evolutionary biology, which, at its core, was still based on population dynamics.

Proponents of evo-devo also adopted a set of additional theoretical concepts in order to account for the patterns of phenotypic evolution and the rapidly emerging insights into the evolution of developmental systems. Within evo-devo these concepts – modularity, evolvability, constraints, homology (in all its modifications and expansions), plasticity, and, more recently, epigenetic and environmental factors – contribute to the theoretical description of the phenomenology of phenotypic evolution. Much of the ongoing work within evo-devo has focused on these aspects of the evolutionary process. But all of it has left the causal structure of evolutionary explanations untouched. At most it offered slight corrections to understandings of the role of natural selection – such as that developmental constraints provide limits to phenotypic variation – or offered additional "non-genetic" causal factors to the genotype–phenotype map – such as environmental factors or physical and chemical properties of cells. As Hall suggested in 2000, the focus here is on synthesis, incorporating additional causal factors and empirical details into the theoretical framework of evolutionary theory.

Developmental evolution, on the other hand, is more radical and aims to transform the explanatory structure of theories of phenotypic evolution. At the SICB symposium we tried to capture this in the title of our paper, "Developmental Evolution as a Mechanistic Science: The Inference from Developmental Mechanisms to Evolutionary Processes" (Wagner *et al.* 2000). Our emphasis here was on developmental mechanism as a primary explanation for processes of phenotypic evolution. But to contextualize this claim, we need to provide some historical background.

Darwinian evolutionary theory started with the recognition of the importance of variation and the realization that heritable variation paired with competition will lead to selection (either natural or artificial). As a consequence the distribution of variants within populations will shift. Since variation has been recognized as an obvious characteristic of natural populations, Darwin and his successors devoted more time to uncovering the causes of inheritance – a necessary condition for natural selection to work – than on trying to understand the causes of variation as such. But this does not mean that they were unaware of the problem. For Darwin the origin of variation was an important part of his theory and he considered developmental mechanisms to be a central part of the explanation of phenotypic variation even though he did not have as many empirical data for this problem as he did for many of the other questions he was investigating.

After Darwin the problem of variation was attracting considerable interest. Many studies dealt with quantitative descriptions of variation, its geographical distribution, and its patterns of inheritance.[25] Developmental considerations about the origin of variation also played a role, albeit smaller, as discussed below. But of all the concepts associated with variation, the one that had the most profound impact on evolutionary theory was the notion of distinct "factors" – soon to be called genes – as the hereditary causes of variation. Genes, and their proposed variants or alleles, could not only account for specific (Mendelian) patters of inheritance, they also gave a precise meaning to one conception of the origin of variation, namely the idea of mutation. If characters are caused by genes, then any variation in such a character could be attributed to a mutation in the underlying gene(s). Soon evidence (1) that mutations can appear spontaneously and (2) that these mutations are inherited in predictable patterns began to accumulate.

As a consequence, evolutionary changes were then described by mutations and the subsequent dynamics of these mutations within populations. The level of genes (or genotypes) and population and quantitative genetics (in both mathematical and experimental versions) became the primary focus of evolutionary theory. Early on, geneticists and evolutionary biologists realized that not all characters follow simple patterns of inheritance. A number of theoretical ideas were introduced to account for these increasing complexities including pleiotropy, epistasis, penetrance, expressivity, and any number of models of multi-genic patterns of inheritance.[26] Common to all of these concepts has been the theoretical assumption that in the context of phenotypic evolution mutations, in whatever shape or form, are a sufficient explanation and that ultimately phenotypic evolution can be described at the level of the genotype.

Once this focus on the genotype had been established – and it is the core assumption of all models of population and quantitative genetics,

[25] See for example Bateson 1992.   [26] See Laubichler and Sarkar 2002.

which are, in turn, the explanatory basis of evolutionary theory from the Modern Synthesis onward – the remaining problem was to fill in the gaps in the genotype–phenotype map. And this is, as we have seen, the *raison d'être* of evo-devo, to provide this missing piece of the evolutionary synthesis. This, at least, is the semi-teleological and triumphalist narrative that has become the party line of much of evolutionary biology and its various popularizers.

Even though we were only able to provide a sketchy description here and not a complete or adequate historical account, any historian or philosopher of biology should now be quite skeptical and ask: really, is this all that has been going on in twentieth-century evolutionary biology? Is evolutionary theory indeed approaching the kind of completion that the proponents of evo-devo seem to advocate? And what would that mean? Might it be the triumph of the Darwinian Method that Michael Ruse and his peers were discussing several decades ago?

Well, the tensions that were on display during the 2000 SICB meeting clearly suggest otherwise. So we need to ask, what are the conceptual antecedents for developmental evolution and what is meant by mechanistic science? Where does the inspiration for the possible revolution that Hall and others associate with developmental evolution come from? This brings us to proportionally much less understood or studied trajectories within the history of post-Darwinian biology. All we can do here is to sketch one alternative that has focused on the role of developmental mechanisms in explanations of phenotypic evolution.

We have already indicated that Darwin argued that developmental processes can provide an explanation of the origin of phenotypic variation, even though, as he put it, "our ignorance" on these matters is "profound." In subsequent decades experimental cell biologists and embryologists learned a lot about these developmental processes and how they bring about organismal form. Some, such as Theodor Boveri or Thomas Hunt Morgan, were clearly also thinking about what role these insights can play in understanding evolution. But their take on the evolutionary process was different. They did not focus primarily on the actions of natural selection; rather they were concerned with the mechanistic, i.e. developmental basis for phenotypic transformations: how can understanding development help us understand the diversity of organismal forms? In other words, their evolutionary focus was on the origin or the generation of variation and not on the relative success of existing variants.[27]

---

[27] See Laubichler and Maienschein 2013.

This focus was in part conceptual, in part also a consequence of their methods of experimentation, which focused on specific controllable interventions to reveal the causal chains that built organismal phenotypes. In light of this methodological separation we can then understand Boveri's claim that the holy grail of an experimental understanding of evolution would be "to transform one organism in front of our yes into another."[28] Not by generations of selective breeding, but by experimental manipulation of the developmental processes that ultimately generate phenotypes in the first place. Much of this was beyond the technical reach of early twentieth-century biology. Boveri, for example, was keenly aware of the substantial challenges that such an experimental and mechanistic approach to problems of phenotypic evolution entailed, and when he was invited to design the Kaiser Wilhelm Institute for Biology he used that opportunity to put together an institute that would be capable of contribution to this agenda. Even though illness prevented him from actually taking up that post himself, with some modifications his plan was implemented and the institute did become for decades a major hub for work in that tradition.

The major conceptual idea behind Boveri's idea that it should be possible to "transform an organism into another" was the notion that development is controlled by a complex regulatory system of anlagen, located in the nucleus, but interacting also with the cytoplasm. The idea here is relatively simple. As cells differentiate in the course of development and become increasingly specialized in a coordinated fashion, the expression of the hereditary material, which causes differentiation, must be highly regulated. Many difficult and elegant observations, from his famous dispermy experiments to the studies of chromosomal diminution in *Ascaris*, contributed to these insights (as well as the chromosomal theory of inheritance). But conceptually the situation is relatively clear. Between the totipotency of the fertilized egg and the differentiated cell types, the expression of the hereditary substances must be regulated. The question was, how.

Morgan, for instance, argued that this could be accomplished through interactions with the cytoplasm and the environment. Boveri and others emphasized that the structure and composition of the cytoplasm at one time is the product of previous actions of the hereditary, i.e. nuclear, substance and focused more on the regulatory properties of the hereditary system as such. In any case, clear experimental evidence was difficult to come by.[29] Individual observations and experimental systems, such as those by

---

[28] Boveri 1906.
[29] Boveri 1906 and Laubichler and Davidson 2008.

Goldschmidt and Kühn, Goldschmidt's successor at the Kaiser Wilhelm Institute, solved some small pieces of the puzzle by revealing some small details of the structure of the hereditary system, such as the notion of macro- or regulatory mutations as the cause of homeotic transformations or the dissection of single genetic pathways and reaction-chains.[30]

The situation changed with the advent of molecular biology. The discovery of the structure of DNA and subsequently the role of different types of RNA in transcription and translation of the hereditary material provided a concrete molecular foundation for both development and heredity. The original and vague conception of a regulatory system of hereditary anlagen gave way to concrete models of gene regulation, foremost among them the Britten–Davidson model of Gene Regulatory Networks (GRN) from 1969.[31] This model focused specifically on both the developmental and evolutionary implications of what we would now call a regulatory genomic system. It proposed, among other things, that changes/mutations in different elements of the genome will have different phenotypic consequences and that regulatory mutations will be involved in producing major phenotypic transformation (those that have been associated with body-plan features). Conceptually, this idea opens up an avenue that could eventually allow us to realize Boveri's challenge: "to transform an organism into another."

But much needed to happen first. The Britten–Davidson model was a logical argument based on fundamental insights into the notion of developmental evolution and whatever little concrete molecular information was available at the time. Decades and many molecular and genomics revolutions later the logical structure of the model stood the test of time. But now we also have the ability to actually fill in all the concrete details. The results are increasingly complete gene regulatory networks, such as those that controls the first 36 hours of the life of a sea urchin embryo.[32] It is now also widely accepted that a mutation is not a mutation is not a mutation.[33] The conserved nature and role of transcription factors, such as the hox genes, strongly suggests that there is a difference between different types of genetic and phenotypic variation. Small-scale variation in often quantitative traits, such as body size or coloration, is often controlled by single-locus polymorphisms (or a small number of loci), and fits the paradigm of population and quantitative genetics. Larger-scale or body-

[30] Laubichler and Rheinberger 2004.
[31] Britten and Davidson 1969.
[32] Davidson 2011 and Erwin and Davidson 2009.
[33] Carroll 2008 and Davidson 2006.

plan features, on the other hand, are the product of regulatory evolution, i.e. changes in the genomic regulatory systems that control development. The theoretical challenge for evolutionary theory is how to integrate and accommodate these two types of genomic changes. Different perspectives on this problem are also at the heart of the evo-devo vs. developmental evolution divide, also a very rich area of research for historians and philosophers of biology.

But this is not all. Rather than suggesting some ways for this problem to be conceptualized from theoretical, epistemological, or historical perspectives, we want to end this brief historical sketch with a quick look at what the future of developmental evolution will likely bring, which, incidentally will bring us back full circle to the turn of the twentieth century and Boveri. Conceptually, the idea of gene regulatory networks as the mechanistic cause controlling development and evolution offers some tantalizing possibilities to actually "transform organisms into each other," or, at the very least, reconstruct evolutionary transformations.[34] The basic logic is again rather straightforward. If we know the GRN of a lineage that acquired a major evolutionary novelty and if we compare it with the GRN of a related lineage that lacks this new feature, we can investigate what kind of regulatory changes contributed to the emergence of a novelty. This is not easy, but doable. The comparison will thus suggest a hypothesis about the causes of phenotypic evolutionary change. Taking advantage of molecular techniques, it will be, in some cases, possible to engineer the GRN of embryos in the lineage representing the ancestral condition and see if they will express, even in rudiments, the novel feature characteristic of the derived condition. This method of synthetic experimental evolution will (and does) then provide an experimental test for hypotheses about the causes of major phenotypic transformation or novelties, exactly as Boveri envisioned it more than a century ago.

Synthetic experimental evolution is more than just a highly specialized experimental procedure. Conceptually, it does represent the kind of revolution that Brian Hall and others, including us, have been arguing that developmental evolution actually is. We now see, in concrete terms, what it means to have a developmental understanding of phenotypic evolution and how understanding the origin of variation is the starting point of all (future) theories of phenotypic evolution. Furthermore, GRNs can now also be turned into computational models, as they are basically Boolean networks (albeit quite complex ones). This opens up the possibility of not

---

[34] Erwin and Davidson 2009.

just in vivo, but also in silico synthetic experimental evolution and provides us with the computational foundation for a formal theory of phenotypic evolution.[35] Again, we were only able to scratch the surface here, but it should be clear that these developments provide huge employment opportunities for historians and philosophers of biology.

Conclusions

Clearly, developmental evolution represents exciting science, with many discoveries shedding new light on old questions. But single insights, such as recent findings about the developmental hourglass or any number of other results, are only part of the story.[36] The transformative potential of developmental evolution is realized in its conceptual structure and the ability to provide an explanation for the origin of variation in terms of developmental mechanisms anchored by the four-dimensional regulatory genome.

As a consequence, developmental evolution does indeed have the kind of revolutionary conceptual and explanatory structure that we, as well as Hall, suggested more than a decade ago. We have argued here that in the context of developmental evolution the causal structure of evolutionary explanation has shifted from a primacy of population-level dynamics to the primacy of developmental mechanisms and that explaining the origin of variation rather than the fate of variants within populations is the first and most important problem for all theories of phenotypic evolution.

Not that evolutionary dynamics are unimportant, rather they act as a (relatively well-understood) filter for variation and as such account for changes over time. But, as many critics of Darwin have pointed out over the last 150 years, selection by itself does not generate variation, and the assumption that mutations, at whatever constant rate, are all that is needed for natural selection to work is no longer a valid argument in light of what we have learned about the complex links between genotype and phenotype. Rather, an understanding of phenotypic evolution that is grounded in the mechanistic interactions of complex regulatory networks, from genomes to the environment, and that will allow us to devise synthetic experimental approaches, both in vivo and in silico, is now feasible. Evolutionary biology will thus become a mechanistic science and, among many other implications, this will also mean that we will finally transcend

---

[35] Peter *et al.* 2012.
[36] Prud'Homme and Gompel 2010 and Kalinka *et al.* 2010.

the canonical distinction between "proximate" and "ultimate" causes, over which philosophers of biology have spilled so much ink.

Much of what we have sketched here is still in its infancy, but it should be clear that Michael Ruse has been right to wish for a long life. There are many interesting challenges for a historian and/or philosopher of biology in that domain, and yes, today's and tomorrow's evolutionary theory is not your grandfather's evolutionary theory. It is your great-grandfather's again!

REFERENCES

Barlow, Nora (ed.) (1958) *The Autobiography of Charles Darwin 1809–1882. With the Original Omissions Restored.* London: Collins.

Bateson, W. (1992) *Materials for the Study of Variation Treated with Especial Regard to Discontinuity in the Origin of Species.* Baltimore, MD: Johns Hopkins University Press.

Boveri, T. (1906) *Die Organismen als historische Wesen.* Würzburg: Königliche Universitätsdruckerei von H. Stürtz.

Britten, R. J. and E. H. Davidson (1969) "Gene Regulation for Higher Cells: A Theory." *Science* 165 (891): 349–57.

Burian, Richard (2000) "General Introduction to the Symposium on Evolutionary Developmental Biology: Paradigms, Problems, and Prospects." *American Zoologist* 40: 711–17.

Carroll, S. B. (2008) "Evo-Devo and an Expanding Evolutionary Synthesis: A Genetic Theory of Morphological Evolution." *Cell* 134 (1): 25–36.

Darwin, Charles (1859) *On the Origin of Species by Means of Natural Selection, or the Preservation of Favoured Races in the Struggle for Life.* London: John Murray.

Davidson, E. H. (2006) *The Regulatory Genome: Gene Regulatory Networks in Development and Evolution.* Burlington, MA: Academic Press.

(2011) "Evolutionary Bioscience as Regulatory Systems Biology." *Developmental Biology* 357 (1): 35–40.

Erwin, D. H. and E. H. Davidson (2009) "The Evolution of Hierarchical Gene Regulatory Networks." *Nature Reviews Genetics* 10 (2): 141–48.

Gerhart, John and Marc Kirschner (1997) *Cells, Embryos, and Evolution. Toward a Cellular and Developmental Understanding of Phenotypic Variation and Evolutionary Adaptability.* Oxford: Blackwell.

Gilbert, Scott F. (2006) *Developmental Biology*, 8th edn. Sunderland, MA: Sinauer Associates.

Gilbert, Scott. F. and D. Epel (2009) *Ecological Developmental Biology: Integrating Epigenetics, Medicine, and Evolution.* Sunderland, MA: Sinauer Associates.

Gould, Stephen Jay (1977) *Ontogeny and Phylogeny.* Cambridge, MA: Harvard University Press.

Hall, Brian K. (2000) "Evo-Devo or Devo-Evo: Does It Matter?" *Evolution and Development* 2 (4): 177–78.

Hall, Brian K. and Wendy M. Olson (2006) *Keywords and Concepts in Evolutionary Developmental Biology*. Cambridge, MA: Harvard University Press.

Hull, David (1974) *Philosophy of Biological Science*. Englewood Cliffs, NJ: Prentice-Hall.

Huxley, Julian (1943) *Evolution: The Modern Synthesis*. New York: Harper and Brothers.

Kalinka, A. T., K. M. Varga, D. T. Gerrard, S. Preibisch, D. L. Corcoran, J. Jarrells *et al.* (2010) "Gene Expression Divergence Recapitulates the Developmental Hourglass Model." *Nature* 468 (7325): 811–14.

Laubichler, Manfred D. (2009) "Evolutionary Developmental Biology Offers a Significant Challenge to the Neo-Darwinian Paradigm." In F. Ayala and R. Arp (eds.), *Contemporary Debates in Philosophy of Biology*. Oxford: Wiley-Blackwell, pp: 199–212.

Laubichler, Manfred D. and E. H. Davidson (2008) "Boveri's Long Experiment: Sea Urchin Merogons and the Establishment of the Role of Nuclear Chromosomes in Development." *Developmental Biology* 314: 1–11.

Laubichler, Manfred D. and Jane Maienschein (2007) *From Embryology to Evo Devo: A History of Developmental Evolution*. Cambridge, MA: MIT Press.

(eds.) (2009) *Form and Function in Developmental Evolution*. Cambridge University Press.

(2013) "Developmental Evolution." In Michael Ruse (ed.), *The Cambridge Encyclopedia of Darwin and Evolutionary Thought*. Cambridge University Press, pp: 375–83.

Laubichler, Manfred D. and H. J. Rheinberger (2004) "Alfred Kuhn (1885–1968) and Developmental Evolution." *Journal of Experimental Zoology Part B: Molecular and Developmental Evolution* 302B: 103–10.

Laubichler, Manfred D. and S. Sarkar (2002) "Flies, Genes, and Brains: Oskar Vogt, Nicolai Timofeeff-Ressovsky and the Origin of the Concepts of Penetrance and Expressivity in Classical Genetics." In R. Ankeny and L. Parker (eds.), *Medical Genetics. Conceptual Foundations and Classic Questions*. Dordrecht: Kluwer, pp. 63–85.

McClung, Clarence E. (1941) "Evolution of the Germplasm." In M. Demerc *et al.*, *Cytology, Genetics, and Evolution*. Philadelphia: University of Pennsylvania Press.

Mayr, Ernst (1982) *The Growth of Biological Thought. Diversity, Evolution, and Inheritance*. Cambridge, MA: Harvard University Press.

Peter, I. S., E. Faure, and E. H. Davidson (2012) "Predictive Computation of Genomic Logic Processing Functions in Embryonic Development." *Proceedings of the National Academy of Sciences* 109 (41): 16434–42.

Prud'Homme, B. and N. Gompel (2010) "Evolutionary Biology: Genomic Hourglass." *Nature* 468 (7325): 768–69.

Raff, Rudolf A. and Thomas C. Kaufman (1983) *Embryos, Genes, and Evolution. The Developmental-Genetic Basis of Evolutionary Change*. New York: Macmillan.

Raff, Rudolf A. and Elizabeth C. Raff (eds.) (1987) *Development as an Evolutionary Process*. New York: Alan R. Liss.

Ruse, Michael (1973) *The Philosophy of Biology*. London: Hutchinson.
  (2005) "Forty Years a Philosopher of Biology: Why EvoDevo Makes Me Still Excited about My Subject." July 5. www.shelfari.com/groups/17494/discussions/18169/Why-EvoDevo-Makes-Me-Still-Excited-About-My-Subject-%28Michael-Ruse (accessed November 23, 2012).
Wagner, Günter P., Chi-Hua Chiu, and Manfred Laubichler (2000) "Developmental Evolution as a Mechanistic Science: The Inference from Developmental Mechanisms to Evolutionary Processes." *American Zoologist* 40: 819–31.
Wilson, Edmund Beecher (1896) *The Cell in Development and Inheritance*. New York: Macmillan.

PART IV

Function, adaptation, and design

CHAPTER 9

Darwin's cyclopean architect

John Beatty

Darwin was concerned to establish not only that natural selection could and did bring about evolutionary change, but also that selection was the most important – "paramount" – of all evolutionary factors. To these ends he countered William Paley's design analogy with design analogies of his own. To demonstrate the competence of natural selection, he compared it to a breeder. To demonstrate the supreme importance of natural selection, he compared it to an architect.[1] It is the latter issue and analogy that concern me here.

The main point of the architect analogy was to emphasize, more specifically, the importance of natural selection relative to the production of variation. Some of Darwin's otherwise staunchest supporters felt he had, in the *Origin*, given short shrift to variation as a factor in evolutionary change. He answered the objections of Joseph Hooker, Charles Lyell, and Asa Gray – and even the self-proclaimed original discoverer of evolution by natural selection, Patrick Matthew – by reiterating that, of course, variation is essential to the evolutionary process. Nonetheless, just as we attribute responsibility for a highly functional building to the architect/builder, and not to his building materials, so too we should give credit for a well-adapted evolutionary outcome to natural selection and not to the variations on which selection has acted. Darwin rethought and modified the analogy on several occasions. These revisions reflect, in part, his changing understanding of the "chance" or "accidental" character of the production of variation.

My purposes here are as much or more analytic as historical. I am concerned not only with the development of the architect analogy, but also with the alternative conceptions of evolution suggested by the different versions. I am particularly interested in the final version, which is ambiguous in several important respects. Depending on how one clarifies

---

[1] Not to be confused with the giant, one-eyed kind. Please read on.

these ambiguities, the importance of the architect (and natural selection) changes relative to the chancy production of the building materials.

Michael Ruse has long insisted that Darwinian evolutionary theory is a design theory through and through (e.g. Ruse 2004). The centrality of the breeder analogy is enough to make that point, and understandably that's where most scholarly attention has been focused.[2] The architect analogy also merits scrutiny, not just as another design analogy, but for the interestingly different role that it plays in Darwin's thought. I find the analogy puzzling because of the ambiguities I alluded to and will elaborate on. A related puzzle is why Darwin did not just re-purpose the breeder analogy to do what the architect analogy was supposed to do. My chapter is pretty much one long exercise in puzzlement (as opposed to Darwin's "one long argument"). I conclude with an unsatisfying (to me) solution, and a restatement of my perplexity, with hopes that the honoree of this volume can set me straight.[3]

Following the publication of the *Origin*, Darwin took the opportunity to clear up some issues left unclear in the book, including his position on the importance of selection relative to the production of variation. As he explained in correspondence with Hooker:

> The following metaphor gives good view of my notion of relative importance of Variability & Selection. – Squared stones, bricks or timber are indispensable for construction of a building; & their Nature will to certain extent influence character of building, but selection I look at, as the architect; & in admiring a well-contrived or splendid building one speaks of the architect alone & not of the brick-maker. (Darwin to Hooker, June 12, 1860, in Darwin 1993, 252; emphasis added)

And writing to Lyell:

> As squared stone, or bricks, or timber, are the indispensable materials for a building, and influence its character, so is variability not only indispensable but influential. Yet in the same manner as the architect is the all important

---

[2] Among Ruse's many contributions to this issue are his influential analysis of the genesis and development of Darwin's artificial-natural selection analogy (1975a), and his absolutely pathbreaking account of the broader philosophical context of appeals to analogy in the mid nineteenth century (1975b and 1979, ch. 3). These are some of his earlier works. But analogies and metaphors continue to be central to Michael's work. For an interesting recent perspective on the breeder analogy, which also includes a nice overview of the literature, see Burnet (2009).

[3] Extended discussions of the architect analogy can be found in Gayon (1998, ch. 1), Beatty (2010), Lennox (2010), Ruse (2010), and especially Ricardo Noguera-Solano's 2013 "Darwin and the Architect Metaphor," which I found online. Since preparing the first draft of this chapter, I learned (thanks to Jonathan Hodge) of Noguera-Solano in press, which is excellent.

person in a building, so is selection with organic bodies. (Darwin to Lyell, June 14, 1860, in Darwin 1993, 254; emphasis added)

In other words, you can't build a building without building materials, but merely having cut stones, bricks, and timber at hand does not guarantee that anything good will come of it. The materials might sit there forever, just as they were delivered. Or they might be put together in a completely dysfunctional way (a building with no entrance). It is the architect who gets credit for "a well-contrived or splendid" building. Similarly, having variation at hand does not guarantee a well-adapted outcome. That is the job of natural selection. Nonetheless, in this version of the analogy, the building materials are not only "indispensable," but depending on how the materials are formed they may be "influential" for the outcome.

In Darwin's next version of the analogy – this one prompted by a query/ complaint from Matthew – the architect's building materials were no longer manufactured in any way appropriate to the task (e.g. no "squared stones"). Instead, the architect had to make do with stones fallen from a cliff that came in an "infinitude of shapes":

> Fragments of rock fallen from a lofty precipice assume an infinitude of shapes – these shapes being due to the nature of the rock, the law of gravity &c – by merely selecting the well-shaped stones & rejecting the ill-shaped an architect called Nat. Selection could make many & various noble build-ings. (Emma Darwin, for Charles, to Matthew, November 21, 1863, in Darwin 1999, 674)[4]

With the brick-maker, the stone-cutter, and the lumber mill operator now out of the picture, the architect clearly gets even more credit for the outcome. And by analogy, natural selection gets even more credit for the outcomes of evolution if the variations selected are not produced to any specification whatever. Of course there can be no building without build-ing materials, and no evolution by natural selection without variation. But other than that trivial sort of importance, the production of materials and organic variations would not seem to have any influence on the architec-tural or evolutionary outcome.

In the final published, and most articulate, version of the analogy, the building materials were once again shaped and sized entirely by their fall, without any tailoring to the task. But this time Darwin further elaborated. Whatever the cause of a particular stone's size and shape, he insisted, it has nothing to do with whether the architect would subsequently make use of

---

[4] I first learned of Darwin's response to Matthew on reading Noguera-Solano's nice treatment of the architect analogy. See the previous footnote.

it. Similarly, whatever the cause of a new variation, it has nothing to do with whether such a variation would be naturally selected. The production of variation, like the production of the architect's building materials, is a matter of "accident" or chance. This understanding of "chance" variation goes beyond the interpretation that Darwin had offered in the *Origin*. There Darwin attributed the appearance of new traits to "chance" only in order to acknowledge his (and others') ignorance as to when, where, and why they arise (1859, 131). He used the published version of the architect analogy to introduce an extended notion of chance variation, this one not entirely based on ignorance: again, whatever the causes of new variation, they have nothing to do with whether the variation will be selected. This has since become the standard understanding of "chance variation" (e.g. "Mutation is random in [the sense] that the chance that a specific mutation will occur is not affected by how useful that mutation would be" Futuyma 1986, 78). Darwin originally articulated this notion, with the help of the architect analogy, in order to defend the overriding importance of natural selection:

> Let an architect be compelled to build an edifice with uncut stones, fallen from a precipice. The shape of each fragment may be called accidental; yet the shape of each has been determined by the force of gravity, the nature of the rock, and the slope of the precipice, – events and circumstances, all of which depend on natural laws; but there is no relation between these laws and the purpose for which each fragment is used by the builder. In the same manner the variations of each creature are determined by fixed and immutable laws; but these bear no relation to the living structure which is slowly built up through the power of selection, whether this be natural or artificial selection. If our architect succeeded in rearing a noble edifice, using the rough wedgeshaped fragments for the arches, the longer stones for the lintels, and so forth, we should admire his skill even in a higher degree than if he had used stones shaped for the purpose. So it is with selection, whether applied by man or by nature; for though variability is indispensably necessary, yet, when we look at some highly complex and excellently adapted organism, variability sinks to a quite subordinate position in importance in comparison with selection, in the same manner as the shape of each fragment used by our supposed architect is unimportant in comparison with his skill. (Darwin 1868, vol. II, 248–49; emphasis added)

By emphasizing the chance production of variation, in the sense he expressed, Darwin sought to further undermine the importance of variation relative to selection in the evolution of well-adapted species. If the causes of variation do not bias it toward (or away) from the direction of selection, then the direction of evolution by natural selection would seem to be unaided (and unimpeded) by the course of variation, and thus natural selection would seem to be solely responsible for the outcome.

Darwin surely had in mind the primitive "cyclopean" form of architecture, as described for example in this period piece:

> When men began to assemble themselves in society, and to occupy fixed habitations, the first great work on which to employ their common exertions would be, surrounding the space covered by their huts by a mound or wall, in order to keep out wild beasts and their still more dangerous human enemies. With this view, the most favourable situation that could be chosen would be a detached rocky hill, of moderate height, flat-topped, and having its flanks more or less protected by inaccessible precipices. On the more gently sloping sides a wall would be raised by collecting the largest blocks lying about, and laying them on one another; at the same time so *adapting* their irregular surfaces as to leave between them the least possible spaces, and filling up these spaces with smaller pieces of stone. Walls of this kind, if skillfully built, and with blocks of large dimensions, even if not united into one mass by the use of cement, oppose, by the mere magnitude and weight of their ingredients, great impediments to disintegration, either from natural causes or external violence. (Aikin 1841, 67, my emphasis)

And by William Whewell (that scholar of architecture [1835] and much else):

> We are reminded of that cyclopean architecture in which each stone, as it occurs, is, with wonderful ingenuity, and with the least possible alteration of its form, shaped so as to fit its place in a solid and lasting edifice. (Whewell 1866, vol. II, 522)

As the first author explains, the cyclopean architect does the "adapting," placing stones appropriately, according to whatever shapes they assume. Whewell allows that the cyclopean architect "with wonderful ingenuity" fits each stone "as it occurs" into place with "the least possible alteration of its form." The style was familiar to Darwin, who used it, e.g. to describe a reef that "when beheld at low water might be mistaken for an artificial breakwater, erected by cyclopean workmen" (Darwin 1874, 266; also 1845, 37). As the first quotation suggests, and Darwin's confirms, the cyclopean style was commonly associated with walls, as in Figure 9.1.

But there were also cyclopean dwellings, monuments, and other structures that had withstood the test of time (more or less). Clearly, "this tedious and difficult method of construction was not adopted from want of architectural skill," rather, very durable structures could be built in this way (Keane 1867, 181).

The bridge over the ravine in Figure 9.2 is a good example of cyclopean architecture. The wedge-shaped keystone in the arch over the culvert was

Figure 9.1 A cyclopean wall

not formed specifically to play that role. Had it been, and had all the other stones been produced specifically for the places they occupy (perhaps with labels indicating which goes where), then the builders would not deserve much applause. Instead, we would be in awe of the process by which the

Figure 9.2 A cyclopean arch bridge

building materials were produced. If, on the other hand, stones of those sizes and shapes just happened to be available, and were put to those uses by the builders – who would have made do with stones of other sizes and shapes had these not been available – then the builders would deserve much greater credit for the outcome.

It is a neat analogy. Like the breeder analogy, it involves comparing natural selection to a familiar causal agency. And of course, as in the breeder analogy, the familiar agency is an intelligent designer. Indeed, one wonders whether Darwin could have re-purposed the breeder analogy to teach the same lesson as the architect analogy, and if so why he did not. The breeder analogy would seem to work. Pigeons with extra tail feathers are not born in order to be selected by fantail pigeon breeders. Rather, fantail pigeon breeders work with the variations that just happen to arise in their stock. Etc., etc. I will return to this issue later.

Again, the architecture analogy has some appeal. But there are ambiguities at the heart of the analogy. And depending on how these are clarified, the chance production of building materials (or organic variation) can assume greater or lesser importance relative to the builder (or natural selection).[5]

---

[5] The ambiguities in question do not have to do with whether natural selection is intelligent, stands upright, has opposable thumbs, or anything like that. I am concerned here with ambiguities that are central to the analogy.

The first ambiguity has to do with the functionality vs. the specific form of the end-product. Is the outcome simply "well contrived" vs. poorly constructed? Or is it one particular, well-contrived structure rather than another that has a different configuration but is equally well built?

To see why this matters, consider a second ambiguity. In the next-to-last version of the analogy, Darwin specifies that the building materials take an "infinitude of shapes," by which I take it he meant great numbers of every size and shape. In the final version, however, he says nothing about the range of variation and the numbers of each variant produced. But this is an important difference. Imagine multiple architects with the same general plans and the same training and competence. Imagine also that each builder has an enormous supply of fallen stones to work with – in every size and shape. From these similar provisions, similarly intentioned and similarly rehearsed architects would fashion very similar structures. Any significant differences would lead us to wonder whether they had the same basic idea in mind after all, or whether they really had the same background and skills. But now imagine that they have quite limited supplies. If the sizes and shapes of the available stones are a matter of chance in the sense Darwin articulated, then the architects will not have the same shapes and sizes of stones, in the same numbers. Each would do his best with the supplies at hand. A builder with a sufficient number of wedge-shaped stones might build an arch bridge, while a builder with a preponderance of long and flat stones might build a clapper bridge instead (see Figure 9.3).

All of the bridges would be functional, and for that the builders would deserve credit. But the functionality would be achieved in possibly quite different ways, and to make sense of that we would need to invoke differences in the materials rather than differences in the builders' initial intentions or competence.

A third ambiguity in the analogy concerns chronology. Whatever the range of materials available, are they all available at the beginning, when the architects begin their work? Or do the architects work with stones as they fall, in the order in which they are formed?[6] When every size and shape is available to the architects from the beginning, we expect very similar results from their labors. But now imagine that the stones appear sequentially with some time in between. Each architect inspects every new stone that appears at his site, and then either incorporates it if it is useful,

---

[6] This may sound like splitting hairs. But as I will explain shortly, the importance of this difference was certainly on Darwin's mind at the time.

Figure 9.3 A cyclopean clapper bridge

or discards it. Discarding materials whose current incorporation would prove dysfunctional, but that might prove useful later, might seem unwise. But if the architects are supposed to work in a way analogous to natural selection, then they're going to have to eliminate as well as accumulate.

Of course, each architect builds from the bottom up. Let's stick with bridge-building. Suppose the initial stones to appear at one site are coconut-sized and irregular. Nonetheless, some of them can be fitted together to form a decent foundation, and the builder begins this way, not knowing if or when something bigger and more rectangular will become available. As it happens, a large cubic stone that would have provided a more stable base appears only after the structure is well underway, and has to be discarded. Wedge-shaped stones that would have been well suited for archways appear too early in the sequence to be useful in that role, and do not reappear when they could be used to complete the openings. Instead a large, flat stone appears and is used to span a gap. And then another. A serviceable clapper bridge goes up. At another site the construction is delayed near the end, because no appropriate finishing materials appear, until finally a wedge-shaped stone appears, and later another. And an arch bridge is completed. Again, all of the completed bridges would be functional, thanks to the architects' skills. But their functionality would be

realized in different ways that do not reflect differences in their builders'
capabilities but rather the chance differences in the order in which materials were produced.

To recap, the architect analogy can be made to speak more positively in
favor of the importance of the chance production of the building materials (and by analogy the chance production of organic variation) by focusing not just on how functional the building is, but rather on the specific
ways in which its functionality was achieved (and by analogy not just on
whether an evolutionary outcome is well adapted, but rather why evolution by natural selection resulted in this well-adapted outcome rather than
another). The chance production of the building materials (and organic
variation) assumes still greater importance if the variability of the materials is restricted, and/or if the materials arise and are employed serially.

    This would not have been news to Darwin, which makes the ambiguity
more perplexing. Indeed, the importance of variation when restricted or
serially introduced was part of Gray's complaint about Darwin overlook-
ing the importance of variation, and Gray more than anyone prompted
the final version of the architect analogy.

    Gray had been happy to accept evolution by natural selection, as long
as it was understood to be God's way of making new species. But Gray
believed that some sort of direction was required in order to get the species that God intended – most notably humans. Accordingly, he proposed
that God has directly or indirectly provided just the right variations at
just the right times in just the right lineages, to be accumulated by natural
selection, thus leading to the evolution of exactly the forms of life He had
in mind. Gray had no problem with Darwin's initial appeal to "chance"
when discussing the production of new variations, since that was only to
acknowledge ignorance as to how precisely God steers the course of variation. Nonetheless, he advised Darwin to assume that somehow "variation has been led along certain beneficial lines" (Gray 1860, 414; see also
Lennox 2010).

    But the notion of chance variation associated with the final version of
the architect analogy precluded the possibility of God providing precisely
the right variations to guarantee preordained outcomes. Immediately following his presentation of the analogy, Darwin turned to Gray's suggestion, which he answered: "However much we may wish it, we can hardly
follow Professor Asa Gray in his belief 'that variation has been led along
certain beneficial lines'" (1868, vol. II, 432). Darwin then acknowledged
what Gray feared most: if variation is indeed a matter of chance, then the

only thing guaranteed is that the outcomes of evolution by natural selection will be well adapted; there is no guarantee that evolution by natural selection will result in humans or any other particular species (1868, vol. II, 430–32).

Which is tantamount to attributing humans not only to natural selection (the architect), but also to the happenstance course of variation upon which selection has acted (the building materials, in the order they were produced). If the divinely ordained course of variation would have been considered crucial to the evolution of humankind, then so too would be the happenstance course of variation.

It was precisely the role of variation in bringing about a particular outcome (humans), rather than just a well-adapted outcome, that concerned Gray. So why, in response to Gray, did Darwin leave the architect analogy ambiguous in this respect?

Darwin himself had focused on the ways in which particular outcomes of evolution can be due to particular, happenstance courses of variation in his book on orchid evolution, published immediately after the (first edition of) the *Origin*. It is interesting, if perplexing, that this illustration of the significance of chance variation was written while Darwin was in the process of articulating and revising the architect analogy, the whole point of which was to demonstrate the "quite subordinate importance" of chance variation. Before addressing these seemingly contradictory attitudes toward the importance of chance variation, it is worth considering the orchid story in order to see how Darwin put chance variation to use in explaining the particular outcomes of evolution by natural selection.

In *The Various Contrivances by Which Orchids Are Fertilised by Insects* (1862, 1877), Darwin sought to establish that the wide variety of orchid flowers all serve the same adaptive end, namely to inhibit self-fertilization and facilitate cross-pollination, but that they accomplish this in vastly different ways: "In my examination of Orchids, hardly any fact has struck me so much as the endless diversities of structure ... for gaining the very same end, namely, the fertilisation of one flower by the pollen from another plant" (1862, 348; 1877, 284). More specifically, orchid flowers employ a wide variety of mechanisms in order to conscript flying insects to transport pollen from one plant to another. These different "contrivances" have evolved, Darwin believed, under virtually the same environmental circumstances, e.g. the same range of available insects. Sometimes one part of the flower has been modified to entice insects in the vicinity, by mimicry or by scent; sometimes another part has been modified to do the same job. Once the insects have arrived, the pollen has to be attached. Some flowers have been constructed so as to catapult pollen at the visiting insects; some catapult the insects against the pollen; some simply induce the visitors to travel past and brush up against the pollen. Etc., etc. Thus cross-pollination has been accomplished in very different ways, the different outcomes being due in large measure to natural selection acting on chance differences in flower morphology among different lineages. Think of multiple architects with similar capabilities and training being given the same task, but somewhat different building materials.

An example that struck Darwin as paradigmatic involved the position of the so-called "labellum" petal, which in most fully formed orchid flowers is the lowermost of the three petals. In that position, it often serves as a landing pad for pollinators. But interestingly, the labellum actually arrives at that position through a 180° twisting of the flower's stem as the flower develops. Darwin reckoned that the position of the labellum in the ancestral orchid had been uppermost, presumably on the grounds that this is also the original position in development, and assuming more generally that the order of development reflects the order of ancestry. He understood the now typical, lowermost position of the labellum to be an outcome of evolution by natural selection of the more twisted variations that had happened, by chance, to arise (Darwin 1862, 349–50; 1877, 284–85).

But Darwin was especially intrigued by cases where the labellum had resumed its uppermost position, which in some cases had resulted from the selection of less and less twisted variations, until there was no twist at all, and in other cases had come about as the result of selection for more and more twisted forms. Flowers of the latter sort twist a full 360 degrees to resume their starting position! As Darwin described the situation:

> [I]n many Orchids the [flower stalk] becomes for a period twisted, causing the labellum to assume the position of a lower petal, so that insects can easily visit the flower; but ... it might be advantageous to the plant that the labellum should resume its normal position on the upper side of the flower, as is actually the case with *Malaxis paludosa*, and some species of Catasetum, &c. This change, it is obvious, might be simply effected by the continued selection of varieties which had their [stalks] less and less twisted, but if the plant only afforded variations with the [stalk] more twisted, the same end could be attained by the selection of such variations, until the flower was turned completely on its axis. This seems to have actually occurred with *Malaxis paludosa*, for the labellum has acquired its present upward position by the [stalk] being twisted twice as much as is usual. (Darwin 1877, 284–85; emphasis added)

So it had apparently become advantageous for *Malaxis* and some species of *Catasetum* to have their labellae uppermost. But due to differences in the variations that chanced to occur in the two lineages, evolution by natural selection had resulted in very different means of serving this end: a 360° twist in the first case, and no twist in the second. Note how restrictions on variation, and the timing of variation, contributed to the difference in outcomes.

Darwin invoked basically the same process – evolution by natural selection of chance differences in variation – to account for the endless diversity of floral morphologies among orchids, with one additional factor. Divergence due to selection of chance differences in variation was, for Darwin, compounded by what we would today call "epistasis": a common situation in which the advantage afforded by one trait depends on what other traits are present. This corresponds to the situation in the architect analogy where the usefulness of a particular size and shape of stone depends on what materials had been incorporated prior to its occurrence. In the case of natural selection, a variation in one trait might prove adaptive only after – not before – evolutionary changes in other traits. For example, Darwin speculated that the adaptive value of having the labellum upright for *Malaxis paludosa* might have been due to some new insects visiting it, but might just as well have been due to a previous change in the shape of the labellum or a previous change in the other petals – a change that was in itself useful for facilitating cross-pollination, but that would work even better with an upright labellum (1877, 284). However, the fact that the upright labellum was due to a 360° twist in one lineage, and no twist in another, was due to a chance difference in the order of variation in the two lines.

To quote Darwin, summarizing the process leading to the diversity of orchid flowers (here I will allow Darwin to repeat himself somewhat):

> In my examination of Orchids, hardly any fact has struck me so much as the endless diversities of structure ... for gaining the very same end, namely, the fertilization of one flower by pollen from another plant. This fact is to a large extent intelligible on the principle of natural selection. As all the parts of a flower are co-ordinated, if slight variations in any one part were preserved from being beneficial to the plant, then the other parts would generally have to be modified in some corresponding manner. *But these latter parts might not vary at all, or they might not vary in a fitting manner, and these other variations, whatever their nature might be, which tended to bring all the parts into more harmonious action with one another, would be preserved by natural selection.*
> To give a simple illustration ... (1877, 284, my emphasis)

At this point Darwin proceeded to relate the case of the super-twisted *Malaxis paludosa*, which illustrated the point that all orchid flowers owe their particular configurations to the chance range and order of variation upon which selection has acted.

Whether (or at least the extent to which, or the sense in which) the architect analogy demonstrates the major importance of natural selection, and the minor importance of the production of variation, depends on how the analogy is interpreted. If the shapes and sizes of the materials are limited, and/or if the materials are made available (and must be used or discarded) sequentially, then the production of variation can have considerable influence on the outcome – if not on its overall functionality, at least on its configuration.

Recall that at one point in the development of the analogy, Darwin replaced cut stones, baked bricks, and sawed lumber with naturally occurring stones taking an "infinitude of shapes." Presumably, in addition, all those sizes and shapes were available to the architect at each stage of the building process. In this way, Darwin sought to guarantee that the production of variation had no influence on the outcome – even its configuration – other than the trivial "indispensability" of having building materials of some sort.

But even in that case, the process by which the materials are produced would seem to be important. What builder would not be indebted to a producer – divine, human, or geophysical – that supplied every size and shape of material, in abundance, every workday of the year? Surely even the most ungodly cyclopean builder would worship the cliff face that produced such a stockpile.

What breeder would not celebrate such an abundance of riches to draw upon in order to satisfy his utilitarian or aesthetic ends? Is this why it did not occur to Darwin (or why he did not choose) to re-purpose the breeder analogy to demonstrate the paramount importance of selection? That is, it just doesn't work very well for that purpose? Breeders do not have such prodigious variation at hand. And some breeders are more successful than others, not because of their skills but because they have more variation to work with. This again is a matter of the chance production of variation. As Darwin acknowledged:

> [A]s variations manifestly useful or pleasing to man appear only occasionally, the chance of their appearance will be much increased by a large number of individuals being kept; and hence this comes to be of the highest importance to success. On this principle Marshall has remarked, with

respect to the sheep of parts of Yorkshire, that "as they generally belong to
poor people, and are mostly in small lots, they never can be improved."
(Darwin 1859, 41)

To account for differences in the outcomes of artificial selection in terms
of differences in the range and numbers of variation in different stocks is
to suggest that similarities in the range and numbers of variation (guaranteed if variation is abundant in every direction) would lead to similarities
in the outcomes of artificial selection (as long as the breeders in question
were interested in the same outcomes and were equally competent). To
regard limits in variation as important (for explaining differences in outcome) but similarities in variation as unimportant (for explaining similarities in outcome) would be oddly asymmetric.

Similarly, how could this sort of proliferation of variation (were it the
case) not figure prominently in accounts of evolutionary outcomes? After
all, if the existence of both 360° and untwisted orchids is attributable to
chance differences in the production of variation in different lineages, then
having the same extensive, ever-present range of variation in all lineages
would presumably preclude or lessen the chance of such differences. Were
variation so luxuriant and common among related lineages, then instead
of the tremendous diversity of orchid floral morphologies, which Darwin
attributed to chance differences in variation in different lineages, there
might instead have been very little diversity. Wouldn't this – on grounds
of symmetry – call for an explanation in terms of there being so much
overlap in the variation produced in each line?

I fail to see any good reason to accord only minor evolutionary importance to the production of variation, relative to the contribution of natural selection. The architect analogy doesn't help me to see otherwise. I'm
further perplexed by the fact that the author of the architect analogy himself attributed considerable evolutionary influence to chance differences in
the production of variation. Again, according to Darwin, natural selection
acting on the happenstance production of variation led to the wonderful
diversity of orchid flowers. Doesn't that make the general happenstance
production of variation important? And natural selection acting upon a
particular happenstance course of variation led to humankind. Doesn't
that make the particular happenstance course of variation important?

I have only a half-hearted suggestion as to why Darwin was so concerned to emphasize the all-importance of natural selection. It has to do
with the question whether the architect analogy concerns merely the func-
tionality/adaptedness of the outcome, or also its particular configuration.
It may be that Darwin was far more concerned about the former than the

latter. For example, he unquestionably emphasized the chance production of variation in accounting for the many different ways in which orchid flowers contribute to cross-pollination. And yet what may have concerned him most was the common high degree of adaptedness of those otherwise diverse evolutionary outcomes. That the influence of natural selection could be demonstrated and not just hypothesized, in every part of every flower he investigated, was surprising even to Darwin, and hence all the more delightful. As he explained his orchid findings to Hooker:

> You speak of adaptation being rarely visible, though present in plants. I have just recently been looking at the common Orchis, and I declare I think its adaptations in every part of the flower quite as beautiful and plain, or even more beautiful than in the Woodpecker [whose various adaptations are so transparent]. (Darwin to Hooker, June 5, 1860, in Darwin 1993, 238)

And as he explained the take-home message of his book manuscript to his publisher, "the chief object is to show the perfection of the many contrivances in Orchids" (Darwin to John Murray, September 21, 1861, in Darwin 1994, 273).

But what makes general adaptedness more telling or weighty or impressive than the diverse manifestations of that adaptedness? From my point of view, nothing. But perhaps I am exaggerating the diversity of the outcomes. After all, similarly experienced architects at work on a common problem – say crossing a culvert – will build something bridge-like, not dwelling-like, while architects that need dwellings will build something dwelling-like rather than bridge-like. Perhaps most people (including Darwin), when shown an arch bridge and a clapper bridge, would first focus on the similarities rather than the differences: "Excellent bridges!" Similarly, while orchid flowers can be very different, they are all (according to Darwin) cross-pollination devices. And to be even more specific, while orchids that twist 360° are very different from those that twist not at all, they both end up with their labellum petal on top, where it is supposedly most useful.

I keep focusing on the differences (360° – holy cow!). Maybe I'm being one-eyed about this. Michael Ruse, what does it look like to you?

REFERENCES

Aikin, A. (1841) *Illustrations of Arts and Manufactures*. London: John van Voorst.
Beatty, John (2006) "Chance Variation: Darwin on Orchids." *Philosophy of Science* 75: 629–41.

(2010) "Reconsidering the Importance of Chance Variation." In Gerd Müller and Massimo Pigliucci (eds.), *Evolution: The Extended Synthesis*. Cambridge, MA: MIT Press, pp. 21–44.

Burnet, D. G. (2009) "Savage Selection: Analogy and Elision in *On the Origin of Species*." *Endeavor* 33: 120–25.

Darwin, C. (1845) *Journal of Researches into the Natural History and Geology of the Countries Visited during the Voyage of H.M.S. "Beagle" round the World, under the Command of Capt. Fitz Roy, R.N.*, 2nd edn. London: John Murray.

(1859) *On the Origin of Species by Means of Natural Selection, or the Preservation of Favoured Races in the Struggle for Life*. London: John Murray.

(1862) *On the Various Contrivances by which British and Foreign Orchids are Fertilised by Insects*. London: John Murray.

(1868) *The Variation of Animals and Plants under Domestication*, 2 vols. London: John Murray.

(1874) *The Structure and Distribution of Coral Reefs*, 2nd edn. London: Smith Elder & Co.

(1877) *The Various Contrivances by which Orchids are Fertilised by Insects*, 2nd edn. London: John Murray.

(1993) *The Correspondence of Charles Darwin, 1860*, ed. F. Burckhardt *et al.*, vol. VIII. Cambridge University Press.

(1994) *The Correspondence of Charles Darwin, 1861*, ed. F. Burckhardt *et al.*, vol. IX. Cambridge University Press.

(1999) *The Correspondence of Charles Darwin, 1863*, ed. F. Burckhardt *et al.*, vol. XI. Cambridge University Press.

Futuyma, Douglas J. (1986) *Evolutionary Biology*. Sunderland, MA: Sinauer Associates.

Gayon, J. (1998) *Darwinism's Struggle for Survival: Heredity and the Hypothesis of Natural Selection*. Cambridge University Press.

Gray, A. (1860) "Darwin and his Reviewers." *Atlantic Monthly* 6 (October): 406–25.

Keane, M. (1867) *The Towers and Temples of Ancient Ireland*. Dublin: Hodges, Smith and Co.

Lennox, J. G. (2010) "The Darwin/Gray Correspondence 1857–1869: An Intelligent Discussion about Chance and Design." *Perspectives on Science* 18: 456–79.

Noguera-Solano, Ricardo (2013) "The Metaphor of the Architect in Darwin: Chance and Free Will." *Zygon: Journal of Religion and Science* 48(4): 859–74.

Ruse, M. (1975a) "Charles Darwin and Artificial Selection." *Journal of the History of Ideas* 36: 339–50.

(1975b) "Darwin's Debt to Philosophy: An Examination of the Influence of the Philosophical Ideas of John F. W. Herschel and William Whewell on the Development of Charles Darwin's Theory of Evolution." *Studies in History and Philosophy of Science* 6: 159–81.

(1979) *The Darwinian Revolution: Science Red in Tooth and Claw*. University of Chicago Press.

202

(2004) *Darwin and Design: Does Evolution Have a Purpose?* Cambridge, MA: Harvard University Press.

(2010) "Form and Function: On Biology and Buildings." *Center: Architecture & Design in America* 15: 94–111.

Whewell, W. (1833) *Astronomy and General Physics, Considered with Reference to Natural Theology*. London: William Pickering.

(1835) *Architectural Notes on German Churches*. Cambridge: Longman and Co.

(1866) *History of the Inductive Sciences*, 3rd edn., 2 vols. New York: Appleton.

CHAPTER 10

Function and teleology

Denis Walsh

> Nearly all biologists and philosophers agree that biology is fundamentally non-teleological; that is they agree that any biological statement which appears to refer to future causes can in fact be translated, without any loss of meaning, into a statement without any such reference.
>
> Ruse 1971, 87

Introduction

A teleological explanation explains the presence or nature of some activity or entity by appeal to the goal or purposive that it subserves. Biology regularly trades in explanations that, taken at face value, appear to do just that. We say, for example, that the elaborate sliding synovial joint on the pectoral girdle of monitor lizards is for extending the power stroke in locomotion (Jenkins and Goslow 1983), and that B cells secrete insulin in order to facilitate the rapid metabolism of blood sugar, and that the function of a bird's syrinx is to produce complex vocalizations. In offering such explanations, biology is like no other mature natural science. Certainly, physics and chemistry employ nothing that remotely resembles teleology. This has been seen quite generally as something of an embarrassment for biology. Any claim that biology uniquely requires a conception of teleological explanation sounds like special pleading. Worse, teleological explanation smacks of the soft-minded Scholasticism that the Scientific Revolution decisively threw over. Evolutionary biology cannot take its place as a science in good standing, and at the same time offer special dispensation to concepts deemed outré and occult by the other sciences. As Monod (1971)

I started thinking seriously about teleology in biology only after an invitation from Michael Ruse to contribute a chapter on the subject to his *Oxford Handbook of Philosophy of Biology* (Walsh 2007). I thank Michael for the opportunity. This chapter is dedicated to Farish A. Jenkins, Jr., who taught me all I know about biological function.

remarked: "The cornerstone of the scientific method is the postulate that science is objective" (12). The postulate of objectivity involves "the systematic denial that 'true' knowledge can be got at by interpreting phenomena in terms of ... 'purpose'" (12).

And yet, to expunge teleological talk from biology completely would require a massive revision of biologists' practice. So, despite its dubious status, teleology – of the sort that we appear committed to when we talk of 'function' or 'purpose' or traits being 'for' something other – seems to be indispensable to biology. Karen Neander states the problem nicely: "What we principally want to understand is how the biological notion of 'a proper function' can be both teleological and scientifically respectable" (Neander 1991a, 454).

An appealing, and popular, approach to understanding biological teleology is to permit biologists to retain a sanitized form of teleological talk by showing that while it is useful *as talk*, it incurs no odious metaphysical commitments to purposes, goals, norms, or values (Ruse 2003). Michael Ruse has long been one of the most forceful and articulate exponents of this strategy. "The teleology in modern biology is analogical. The organic world seems as if it is designed; therefore we treat it as designed. The *artifact* model is the key to biological teleology" (Ruse 1981, 93). The general idea is that although organisms are not artifacts, they are sufficiently like artifacts that it is useful to speak of them *as though* they are.[1]

> End-directed thinking – teleological thinking – is appropriate in biology because, and only because, organisms seem as if they were manufactured, as if they had been created by an intelligence and put to work. (Ruse 2003, 268)

But this deflationary strategy incurs costs of its own. It must show how this analogical use of teleological concepts evades the charge of implicit commitment to genuine teleology. Here again, Ruse has been one of its most astute defenders.

> The design of organisms is to be understood in terms of their survival and reproduction, as Darwin insisted ... Something is of value because it leads to the end of survival and reproduction, but this survival and reproduction are in turn the reason why it exists. (Ruse 2003, 269)

Selection is a mechanism, we are told, that emulates design but without a designer's intentions. Thanks to Darwin, biology is a wholly mechanistic science just like any other, and has no truck with teleology. David Hull

---

[1] Tim Lewens (2005) provides a recent update to the strategy.

trumpets the ascendancy of Darwin over teleology: "From the point of view of contemporary biology, both vitalism and teleology are stone-cold dead" (1969, 249). And yet, biologists are free to talk as though the biological world were imbued with purpose.

The cornerstone of this rehabilitation of teleological talk in biology is the Etiological Theory of Function. That theory promotes the idea that all uses of teleology in biology can be cast as functional statements of the form 'The function of x in z is to y'. These in turn can be translated into statements that refer to the effects of natural selection in the past. For many years, the Etiological Theory of Function has been taken to be that rarest of things: a philosophical result – a genuine problem that has yielded unconditionally to philosophical analysis.

I'm afraid that this is a delusion. The Etiological Theory of Function is less a triumph of philosophical analysis than a victim of it. In this chapter I argue that the Etiological Theory of Function is the product of three methodological assumptions. I call them 'univocity', 'analysis', and 'mechanism'. Together, they merely create the illusion that biology has sloughed off its commitment to teleology. They do so by distorting the range and variety of explanations to which the concept of biological function contributes. Further, they illicitly disguise genuine teleological explanations as a species of causal explanations. The upshot is that what looks like an elegant eliminative reduction of biological teleology is an impediment to the proper understanding of the place of teleology in evolution.

Ruse's contribution

Given the context, it is appropriate to consider Michael Ruse's contribution to the project of de-teleologizing biology. Ruse is a pioneer. The epigraph to this chapter is the first sentence of Ruse's first foray into the project of naturalizing function statements. The elegant little paper that follows this proclamation is remarkable for at least a couple of reasons.

One is its content. Ruse's opening sentence reveals that he is fully aware of the far-reaching implications of (what later became known as) the Etiological Theory of Function. The project is to expunge teleology from evolutionary biology, while affirming the legitimacy of its many evidently teleological locutions. The motivation is clear too; teleology appears to incur unacceptable metaphysical commitments (in Ruse's case, backward causation). The epigraph also discloses the methodological hallmark of the philosophy of its day. Ruse is undertaking an exercise in analysis, or semantics; we are being offered an account of the *meaning* of teleological talk – a translation scheme.

The other important thing to notice about Ruse's paper is its date, 1971. Ruse is generally given little credit for being in the vanguard of the movement to naturalize biological teleology in this way. Yet this paper sketches the contours of the Etiological Theory of Function two years before Wright's (1973) generalized causal account of functions. It occurs 13 years before Ruth Millikan (1984) demonstrated the application of the etiological approach to naturalized intentionality. It predates the definitive formulation of the etiological theory by a full two decades (Griffiths 1993; Godfrey-Smith 1994). It appeared almost three decades before the two most important collections of essays that celebrate the results of this philosophical project.[2]

The other telling feature of the date is that 1971 was a long time ago. Even in philosophy, 43 years can be a while. It can be an especially long time in a dynamic, empirically driven field like evolutionary biology. Ruse's development of the etiological theory offers us a glimpse into the evolutionary biology of the day and its philosophy. In doing so, it also furnishes us the opportunity to take stock, to ask what has changed. I want to suggest that so much has changed in both evolutionary biology and its philosophy as to render void the motivating idea that we need an analysis of function that eliminates the teleology from evolutionary biology.

The Etiological Theory of Function

The Etiological Theory of Function starts with two suppositions. The first is that all uses of teleological language in biology can be cast as instances of the form: 'The function of x in z is to y', where x is a type of process or structure and z is an organism type, and y is an activity. The second is that such locutions are by their nature explanations. As scientific explanations, these are decidedly odd. They explain the presence or nature of things by citing their effects. Outside of artifacts and actions, this isn't typically done. Furthermore, such explanations appear to have normative implications (Neander 1991a). We can distinguish between the 'proper' function of a trait and various ways it might malfunction. Moreover, not every effect of x is a function. Some of its effects are mere accidents. Curiously x may have y as a function, even if 'x does y' isn't true. Typically, scientific explanations do not support these kinds of distinctions; scientific explanations are non-normative and factive.

---

[2] Allen *et al.* (1998) and Buller (1999). Neither collection reprints Ruse's contribution.

There are two general strategies for dealing with this odd category of biological explanations. The first involves simply denying that functional explanations in biology have these implications (Cummins 1975). For Cummins, functional explanations are causal explanations like any other. The function of a trait is just some effect it might have with respect to some system of which it is a part. Ascribing a function to x explains the effect of x and not the presence of x.

The second, more common strategy takes the apparent implications of functional explanation more seriously. Early exponents of this approach include Nagel (1961) and Hempel (1965). These authors attempt to preserve the wholly distinctive nature of function explanations, while drawing them under the ambit of their DN conception of scientific explanation. Ascribing a function to x explains x's presence, they say, by demonstrating that x is necessary for the fulfillment of some activity y. According to Hempel and Nagel 'the function of kidneys is to purify blood' means in part that 'kidneys are necessary for the purification of the blood'. The fact that blood is purified, then, entails – and thus explains – the presence of kidneys.

The Hempel and Nagel approach simply buckles under the weight of counterexamples (Ruse 1971; Cummins 1975). Kidneys may be *for* purifying the blood, but they are not necessary for doing so. Dialysis machines may have the same effect. Nagel (1961) is aware of these shortcomings and attempts to ameliorate them by adding a condition, that organisms are goal-directed. His analysis of the statement 'The function of chlorophyll is to enable plants to perform photosynthesis' goes as follows:

(1) Chlorophyll is necessary for the performance of photosynthesis in plants.
(2) Plants are goal-directed, that is to say, they are capable of persisting towards some end, despite fluctuations in their environment. (Quoted in Ruse 1971, 87.)

In this context, Ruse's approach to functional locutions, and the amendments that were to follow, are a master stroke. Ruse shares with Hempel and Nagel the conviction that in biology, to ascribe a function to x is to explain x's presence or prevalence. But he explicitly denies that the goal-directedness of organisms invoked by Nagel plays any part in the conditions required for function ascriptions. The explanatorily relevant notion here is not that of goal-directedness, but adaptation. Ruse says:

"The function of x in z is to do y" implies that y is the sort of thing which biologists call an 'adaptation'. Consequently, a functional statement in biology draws attention to the fact that what is under consideration is an adaptation, or something which confers an 'adaptive advantage' on its possessor. (Ruse 1971, 89)

Ruse's analysis does two things. First it definitively denies that the goal-directedness of organisms plays any role in functional explanation, and second it explicitly ties function to the process of evolutionary adaptation. The first of these discharges any lingering commitment to teleology; the second brings function ascriptions under the ambit of evolutionary theory. Having cleared up these confusions, Ruse proceeds with his own analysis.

We are now in a position to state the equivalent non-teleological formulation of the generalized function statement:
"The function of x in z is to do y."
It is,

(i) z does y by using x
(ii) y is an adaptation. (Ruse 1971, 91)

The function of chlorophyll is photosynthesis thus means 'plants (z) photosynthesize (y) by using chlorophyll (x), and photosynthesis (y) is an adaptation'.

Adaptation, in turn, needs a suitably naturalistic account. But here Darwin's theory obliges. *Adaptation* is the process by which natural selection causes individuals to survive and reproduce differentially on the basis of their heritable traits. *An adaptation* is a heritable trait type that has become prevalent in a population because of the positive causal contribution it makes to the survival and/or reproduction of the individuals that possess its tokens.

This is the core of the Etiological Theory of Function. To be sure, there were deferents and epicycles, refinements and amendments to come for the next 20 years or more.[3] But its essence was succinctly captured by Ruse's elegant little paper, way back in 1971.

Methodological assumptions

There are three crucial methodological assumptions embodied in the Etiological Theory of Function. They usually go unarticulated, but making them explicit, I believe, can help expose the weaknesses of this entire approach to naturalizing biological teleology. They are: (i) *univocity*, (ii)

---

[3] These are nicely collated in Allen *et al.* (1998) and Buller (1999).

analysis, and (iii) *mechanism*. It is these methodological assumptions that together lead the etiological theory astray.

Univocity

The various accounts of function generally assume that there is a single set of uses to which we typically put function ascriptions in biology, and a single set of conditions under which their ascription is appropriate. The project of the analysis of functions is to specify these uses and to articulate these conditions.[4] Kitcher articulates the assumption explicitly: "I shall start with the idea that there is some unity of conception that spans attributions of functions across the history of biology and across contemporary ascriptions in biological and non-biological contexts" (1993, 379). Kitcher then proceeds to identify the unifying explanatory idea:

> [S]election lurks in the background as the ultimate source of design, generating a hierarchy of ever more specific selection pressures, and the structures, traits, and behaviors of organisms have functions in virtue of their making a causal contribution to responses to those pressures (1993, 390)

Even amongst the few dissenters to the etiological theory, the assumption of univocity holds. Cummins argues that accounts of the nature of function ascriptions prior to his own have labored under the misapprehension that: "The point of functional characterization in science is to explain the presence of the item (organism, mechanism, process, or whatever) that is functionally characterized" (1975, 741). The Cummins account of function, which is commonly seen as the other principal contender for a theory of functions, is predicated upon the idea that *all functional explanations* do the same thing. They explain how an entity contributes to the activities of interest of a system of which it is a part.

Analysis

The second assumption is that what is called for is an *analysis* of function locutions. Such an analysis attempts to do two things at once. It sets out to provide a translation scheme for functional statements. As Ruse himself says: "In this discussion my particular concern will be with finding the correct non-teleological translation of an important subclass of ... biological statements, namely functional statements" (1971, 87). It does

---

[4] Exceptions to this assumption include Walsh and Ariew (1996) and possibly Godfrey-Smith (1993).

so by specifying an analysans condition, ∅, that is co-extensive with 'The function of x in z is to do *y*'. The analysis then takes ∅ to stand in for the *meaning* of the function statement (Neander 1991b). To be sure, there is some resistance to construing the etiological theory as analysis. Millikan (1989) provides a colorful foil:

> Now I firmly believe that "conceptual analysis" taken as a search for necessary and sufficient conditions for the application of terms ... is a confused program, a philosophical chimera ... a misconceived child of a mistaken view of the nature of language and thought. (Quoted from Allen and Bekoff 1998, 297)

The disagreement between Neander and Millikan is less acute than it appears. It turns on whether an analysis must seek to capture the extension of the pre-theoretic analysandum perfectly. But there is a point of agreement on another traditional feature of analysis: analyses capture the *meanings* of their analysanda. In both Neander's and Millikan's version of the Etiological Theory of Function (as in any other), the objective is to capture the meaning of function ascriptions in non-teleological terms. On the assumption (above) that to ascribe a function to a trait *x* is to explain its presence (or prevalence), then the analysans condition, ∅, must also be one that *explains* the presence (or prevalence) of *x*, which brings us to the third methodological assumption about the nature of explanation.

Mechanism

The etiological theory takes on the particularly modern conviction that to explain is to cite causes. Given that, if ascribing a function to *x* explains *x*'s presence, then the function ascription must cite the cause of *x*'s occurrence. It is commonly assumed that natural selection is a causal mechanism that is particularly well suited to explaining the presence or prevalence of a trait. By promoting individuals by dint of their heritable traits, selection also serves to promote the fixation of those traits in a population. This makes the appeal to natural selection in the past a natural candidate for explaining the presence of traits in a population (Neander 1995).

The early influence of the mechanistic approach to explanations is evident in Ruse's critique of Hempel and Nagel. For Hempel and Nagel, an explanation is an argument. Explanantia explain their explananda by entailing them. The objective, then, behind Nagel's stipulation that traits are *necessary* for the performance of their function was to allow that the successful performance of the function *entails* (and hence explains) the existence of the trait. The 1970s saw a growing dissatisfaction with the idea that explanation is deduction. Many authors (e.g. McCarthy 1977) began to explore the idea that in order to explain an event it is necessary and sufficient to cite its causes. This line of thought culminates in the new mechanist conception of explanation laid out by Wesley Salmon. "Causal processes, causal interactions, and causal laws provide the mechanisms by which the world works; to understand why certain things happen, we need to see how they are produced by these mechanisms" (1984,132). As R. Miller was later to assert: "An explanation is an adequate description of underlying causes helping to bring about the phenomenon to be explained (1987, 60).[5] In this too, Ruse is remarkably ahead of his time. He recognizes – as do the etiological theorists to follow – that to explain the presence or prevalence of a trait, one need only cite the process of selection that caused it. For some adaptive trait, x, its capacity to do y in z causes it to be promoted by selection. That capacity both explains its presence and constitutes its function.

Problems for the etiological theory

There is a battery of well-rehearsed objections to the Etiological Theory of Function. They amount to a complaint that that there are more biological functions than there are etiological functions.[6] Defenders of the Etiological Theory have made amendments intended to allay this worry while retaining the essence of the account (Griffiths 1993; Godfrey-Smith 1994). These adjustments may or may not be successful. However, it seems to me that the standard criticisms fail to pinpoint the real problem for the Etiological Theory of Function. The problem for the etiological theory stems directly from the methodological assumptions outlined above: univocity, analysis, and mechanism. Together these assumptions lead to three errors. First, in miscasting the concept of biological function as exclusively explanatory (and etiological) they underestimate the variety of ways in which the concept is used in biology. Second, in offering an analysis of functional explanations, they conflate the conditions under which an explanation may be used with the content of that explanation. Third, and most important, they prevent us from recognizing a distinctive class of wholly natural teleological explanations, which, it appears, are becoming increasingly important in the understanding of adaptive evolution. I discuss the first two

---

[5] Quoted in Ruben (1990), 211.
[6] See Bigelow and Pargetter (1987), Amundson and Lauder (1994), Walsh (1996), Walsh and Ariew (1996).

complaints in the remainder of this section, and reserve the next section exclusively for the third complaint.

Ascription and explanation

The assumption that function ascriptions are by their nature explanatory, combined with the assumption of analysis, suggests that we are looking for necessary and sufficient ways for offering an explanation of the presence of the entity to which we are ascribing a function. Mechanism further suggests that this explanation must cite the causes of its occurrence. Univocity suggests that whatever account we give to some function ascriptions should apply to all. Taken together, these assumptions obscure the important distinction between function ascription and functional explanation. Function ascriptions are not explanations per se; they attribute a property to an entity (or to entities of a kind), and that's all. Function ascriptions may be *used* in explanations, but not all functional explanations are the same.

The Cummins (1975) approach to function has often been misappropriated, even misrepresented by its proponents, as an analysis of functional *explanation*. It is more appropriately seen as an account of function *ascription*. As such, it is a good one. According to Cummins, we ascribe a function in the context of a functional analysis.[7] In a functional analysis we take a complex system whose workings we wish to understand. We dissect it to see how the parts in question contribute to the overall activity of interest of the complex system. The contribution that a part makes to the activity of the whole is its function. These are the conditions under which we make a function ascription. That goes for *any* kind of function (Walsh and Ariew 1996). Where the function in question is an evolutionary function, we investigate how the trait in question *typically* contributes (or contributed) to survival and reproduction of the individuals that possess (or possessed) it. The simple point is that to ascribe a function to an entity (or type) is to say what causal contribution it makes or made (or typically made) to the activities of interest of some complex system.

Function ascriptions are univocal, but their explanatory uses are multifarious. Often, as Cummins says, we ascribe a function as part of the practice of giving an explanation of what an entity does. We may, for example, explain the role of legume rhizobia in the nutrient cycle by saying that

---

[7] Not to be confused with a conceptual analysis.

they function to harbor nitrogen-fixing bacteria. That function ascription may not explain why they're there, but it explains what they do.

Function ascriptions have further explanatory applications. They may be deployed in historical explanations that account for the presence of a trait type by citing its typical function, that are nevertheless not adaptive explanations. For example, booid snakes have vestigial femurs because *in the past*, booid ancestors had hind legs whose femurs contributed to locomotion. This historical function (at least partly) explains the presence of femurs in current snakes, but in no way implies that femurs in booids contribute in locomotion, or that they are *booid* adaptations.

Then there are adaptive explanations proper. These use function ascriptions too. They explain the presence or prevalence of a trait type in organisms of a particular kind, by citing a particular kind of historical function, namely the typical contribution of tokens of that trait type to the differential survival and reproduction of individuals that possessed the trait in question. This is what the Etiological Theory of Function takes *all* functional ascriptions to mean.

There are, in addition, explanations of the contributions of traits to survival and reproduction that apply even to novel traits that have no history of selection (Amundson and Lauder 1994; Walsh 1996). These are not adaptive explanations; they do not refer to the history of selection. Nevertheless they do exploit function ascriptions.

Finally, there are genuinely teleological explanations in biology (Ayala 1970). For example, the beta cells in my pancreas have just secreted insulin. The explanation is that my blood-sugar levels have just increased and the function of insulin is to reduce excess blood-sugar levels. This is a perfectly good biological explanation. It is not in the least metaphysically suspect. It doesn't trade in any dubious form of backward causation. It explains the presence of an event – the secretion of insulin – by its causal contribution to the attainment of a goal, the maintenance of appropriate blood-sugar levels. The effect of insulin in this system is its function, just as Cummins supposes.

Here an observation of Nagel's (1961, 1977), which, as we saw, was summarily dismissed by the etiological theory, becomes germane. These teleological explanations, like explanations of the activities of the endocrine system, work because organisms are goal-directed entities. Organisms have myriad homeostatic capacities (Ayala 1970). They have a multiplicity of systems able to restore homeostasis by making compensatory changes; the endocrine system is one such. This capacity of the endocrine system underwrites the application of teleological explanations. When an

organism has the ability actively to restore its blood sugar to the appropriate level by secreting insulin, typically the fact that insulin has this function helps explain the organism's having secreted it.

Notice what this explanation *doesn't* say. It doesn't say that insulin secretion is prevalent in humans because in the past it has contributed to survival and reproduction by regulating blood-sugar levels. It doesn't say that
the regulation of blood-sugar levels is an adaptation for which insulin is
useful. That may be true, of course, but it doesn't follow that the explanation of why my beta cells have just secreted insulin *says* this. The explanation is not, after all, the explanation of the prevalence of a type of event in
a type of organism. It is an explanation of the occurrence of a *token* event
in a *token* organism. This is an explanation that uses a function *ascription*, but it is not an adaptive explanation; it is genuinely and irreducibly
teleological.

The upshot is that the assumption that function ascriptions in biology
are univocal, and that the objective of an analysis of function ascriptions
is to offer a translation schema obscures the distinction between function
ascriptions and functional explanations. In doing so it distorts the range
of genuinely explanatory uses to which function ascriptions are applied in
biology.

Content and conditions

The foregoing example serves to highlight a further shortcoming of the
Etiological Theory of Function: the etiological theory conflates the content of a functional explanation with the conditions necessary for its
application. Suppose that my beta cells have the capacity to secrete insulin only because of certain historical conditions. In Ruse's schema, for
this functional claim to be genuinely explanatory it would have to be
true that:

(1) mammals (z) regulate blood sugar (y) by using insulin (z) and
(2) regulating blood sugar (y) is an adaptation.

I'm willing to accept this, at least in this case. These are the conditions
required for me to offer the teleological explanation of the event of my
beta cells just now having secreted insulin. But the etiological theory goes
further than this. It doesn't just articulate a set of necessary conditions, it
also gives us a translation schema; it is intended to capture the *meaning* of
the teleological explanation. As Ruse says, in this chapter's epigraph, the
objective of the analysis is to demonstrate that "any biological statement

which appears to refer to future causes can in fact be translated, without any loss of meaning, into a statement without any such reference".

Unfortunately, to translate the teleological explanation of why my beta cells have just secreted insulin as an etiological explanation of why mammalian beta cells in general have this capacity is completely to distort the content of the genuinely teleological explanation. There is a virtually complete loss of meaning. The teleological explanation does not purport to explain the presence of an event or trait *type* in a *kind* of organism. On the surface, at least, it purports to explain the occurrence of a *token* event in a *token* individual. It does so by appeal to some goal or end that the event that the token event subserves, namely the maintenance of the appropriate blood-sugar levels. The fact that the event in question is conducive to my goal of survival explains why it occurs. It may be a necessary condition on my being able to offer this explanation that the mammalian endocrine system is adapted for such purposes, but to suppose that the explanation *means* that the mammalian endocrine system is an adaptation (inter alia) for regulating blood sugar is simply to conflate the conditions required for a successful explanation with the content of that explanation.[8]

My guess is that the assumptions of mechanism and analysis are jointly to blame here. The supposition that function ascriptions are explanations, and that all functional explanations are mechanistic, leads to a conflation of the conditions required to offer a teleological explanation and the content of those explanations.

Quite apart from the issue of culpability, a more important point emerges. When we take care to mark the distinction between the conditions required for an explanation and the content of that explanation, we find that there are, indeed, genuine teleological explanations in biology, explanations that explain the presence or occurrence of a token event or thing by appeal to some goal that it subserves.

Natural teleology

Having made the concession that there are genuine explanations in biology that cite goals, we immediately face two questions: (i) how do goals explain? and (ii) how important are these explanations for evolutionary biology?

---

[8] I thank Fermin Fulda for discussions on the conflation between the content of an explanation and the conditions required for offering it.

Goals and explanations

Part of the general problem of understanding how goals might explain derives from a difficulty in specifying what goals are. It is sometimes thought that goals are intrinsically normative or evaluative states of affairs (Bedau 1992). That being so would require us to "reckon objective stand-ards of value as part of the natural order" (Bedau 1991, 655). This, of course, is anathema to most naturalists. But no such anti-naturalist concessions are required in order to characterize a goal. A goal is simply an end-state to which a goal-directed system tends. Goal-directedness is an observable, gross behavioral property of a system. A system – like your endocrine, or immune, or thermoregulatory system – manifests goal-directed behavior when it is capable of maintaining a stable end-state through the imple-mentation of responses that are appropriate, under the circumstances, to return the system to its equilibrium, or stable end-state, across a range of conditions. In order to have this capacity, a goal-directed system must have a repertoire – a range of possible responses to various circumstances. Further, this repertoire must be 'biased'; that is to say, the system must have a tendency to implement those elements of its repertoire that are conducive to the attainment or maintenance of the end-state, across a range of circumstances. Given these conditions, the end-state is the goal. If that is so, then no fundamentally normative or evaluative properties are required to specify the nature of goals.

If that is what a goal *is*, then how can goals explain? Here we can bor-row a page from the mechanists' book. Mechanists have a simple answer to the question 'how do causes explain?', or to paraphrase, 'what is the relation between cause and effect such that causes can be used to explain their effects?'. Their answer is 'invariance'. Woodward outlines this con-ception of the relation between cause and effects that underwrites mecha-nistic explanations:

> I understand this in terms of the notion of invariance under interventions. Suppose that X and Y are variables that can take at least two values … The intuitive idea is that an intervention on X with respect to Y is a change in the value of X that changes Y, if at all, only via a route that goes through X and not in some other way. (Woodward 2002, S369–S370)

He later elaborated:

> This sort of counterfactual dependence is required for explanation … the sorts of counterfactuals that matter for purposes of causation and explana-tion are just such counterfactuals that describe how the value of one vari-able would change under interventions that change the value of another.

Thus, as a rough approximation, a necessary and sufficient condition for X to cause Y or to figure in a causal explanation of Y is that the value of X would change under some intervention on X in some background circumstances. (Woodward 2003, 15)[9]

The idea is surprisingly simple. We can explain effects by citing causes because the relation between them is such that were the same causes to occur in similar background conditions, the same effects would occur, and were different causes to occur in similar conditions a different effect would occur. Causes are a counterfactually robust kind of difference-maker (Strevens 2009).

Remarkably, the same kind of relation holds between a goal and the means to its attainment. A goal is a counterfactually robust kind of difference-maker. For any system with goal e_1 that produces an event m_1, that is conducive to e_1 under actual conditions, under different conditions it would produce a different event m_2 such that m_2 would, in those conditions, conduce to e_1. Under similar circumstances, were the system to have had a different goal, e_2, then it would have produced a different event, m_3, conducive to e_2.

The relation between an end (goal) and its means is an invariance relation. It is exactly the obverse of the relation that holds between a cause and its effects. If invariance underwrites the explanatory relation between cause and effect, then it ought to underwrite an analogous relation between goals and means. Just as we can say that a cause explains its effect because were the cause to obtain then so would the effect and were it not to obtain the effect would not either, we can also say that a goal explains its means because were the system in question to have the goal then the means would obtain, and were it not to have the goal the means would not have occurred. In this way, goals explain (Walsh 2012, in press a). Notice that this account of teleological explanation in no way incurs a commitment to backward causation, or to causation by non-actualia. It simply draws on the resources of the most intuitively appealing account of causal explanation – in particular that explanation is underwritten by invariance.

Teleology in evolution

If teleological explanation is permissible in biology, the question arises how broad its brief is. The general supposition is that the significance of

[9] Woodward did not intend his account of causal explanations to extend to an account of teleological explanations. I take it that invariance between X and Y is only *necessary* for X to be part of a *causal* explanation of Y. (See Walsh 2012, in press a, in press b for an extended discussion.)

teleological explanations is limited (Ayala 1970). They apply only to those goal-directed systems, like the endocrine, immune, and thermoregulatory systems, that are specific adaptations to dealing with uncertain circumstances. But, so the common thought goes, teleology is generally unimportant to an understanding of evolution. Increasingly, however, evolutionary biologists are questioning this exclusion of teleology from the standard account of evolution (Newman 2009). Organisms are the paradigm cases of goal-directed systems. By their natures, they are uniquely well set up to deal with uncertain circumstances. Biologists are beginning to recognize that this goal-directedness itself plays a crucial role in making the process of adaptive evolution adaptive.

There are three basic areas in which the significance of goal-directedness for evolution is becoming apparent: (i) the generation of evolutionary novelties, (ii) the fidelity of inheritance, and (iii) the function of genomes. I take these in turn.

Plasticity and novelty

The distinctive feature of organisms is their robustness (Kitano 2004). Robustness is the capacity of an organism to attain or maintain a stable end-state through adaptive responses to perturbations. It consists in the ability to produce a process or structure precisely *because*, under the circumstances, that process or structure is conducive to an organism's survival. Kirschner and Gerhart encapsulate this feature nicely:

> The organism is not robust because it is built in such a manner that it does not buckle under stress. Its robustness stems from a physiology that is adaptive. It stays the same, not because it cannot change but because it compensates for change around it. The secret of the phenotype is dynamic restoration. (Kirschner and Gerhart 2005, 108–9)

This form of robustness is alternatively known as phenotypic plasticity, and is increasingly thought by many biologists to be the fundamental defining feature of living things: "Phenotypic plasticity is a ubiquitous, and probably primal phenomenon of life" (Wagner 2011, 216).

Adaptive phenotypic plasticity has particular significance for development. Plasticity confers on organisms a capacity for accommodating to developmental, environmental, and genetic uncertainties. It appears to be a prerequisite for adaptive evolution (Schlichting and Moczek 2010). "There are architectural requirements for complex systems to be evolvable, which essentially requires the system to be robust against environmental and genetic perturbations" (Kitano 2004, 829). Phenotypic plasticity has two significant roles in adaptive evolution. The first is in the production of novelties : "Responsive phenotype structure is the primary source of novel phenotypes. And it matters little from a developmental point of view whether the recurrent change we call a phenotypic novelty is induced by a mutation or by a factor in the environment" (West-Eberhard 2003, 503). In fact, adaptive phenotypic change, rather than genetic mutation, seems to be the principal source of such novelties (Wagner 2012). "Adaptive phenotypic adjustments to potentially disruptive effects of the novel input exaggerate and accommodate the phenotypic change, *without genetic change*" (West-Eberhard 2005, 613, emphasis in original).

> The second significant role lies in the orchestration of adaptive evolution. The evolution of complex adaptations requires a significant degree of coordination between an organism's various systems. A change in, say, the strength of an appendicular muscle requires a concomitant change in the load-bearing capacity of its associated bone, an increase in vascularization and concomitant changes to the nervous and integumentary system. Phenotypic accommodation permits all these systems to respond to a phenotypic novelty in ways that secure the viability of the organism as a whole. No genetic change is required. If it were, the evolution of complex adaptations might be well-nigh impossible. In contrast to the rapid response produced by plasticity, if the production of newly favored phenotypes requires new mutations, the waiting time for such mutations can be prohibitively long and the probability of subsequent loss through drift can be high. (Pfennig *et al.* 2010, 459–60)

In sum, then, the capacity of organisms to respond to changes and perturbations in ways that preserve their viability is required to *explain* the origin and maintenance of novel phenotypic characters in evolution. Evolution is adaptive, because organisms are adaptive, goal-directed systems. Novel phenotypes, and the accommodations they induce, occur when they do *precisely because* they contribute to the organism's goals of survival and reproduction. We need to invoke the capacity of organisms to pursue goals in order to explain the origin of adaptive novelties.

The high fidelity of inheritance

One of the cornerstones of twentieth-century evolutionary biology is the conviction that the intergenerational stability of phenotype required for cumulative evolution is a function of the unchanging nature of genes or replicators. On this traditional view, genes are by their very natures highly stable and conservative. They are particularly resistant to alteration by any processes downstream of replication and translations. As Jacques Monod avers: "[T]here exists no conceivable mechanism whereby any instruction or piece of information could be transferred to DNA ... Hence the entire system is totally, intensely conservative, locked into itself, utterly impervious to any 'hints' from the outside world" (Monod 1971, 110). Yet it is becoming evident that the high fidelity of phenotypic inheritance owes itself very largely to two goal-directed capacities of organisms. The first is the capacity to detect, respond to, and repair DNA lesions that would otherwise be lethal. The second is the capacity of organisms' developmental systems reliably to produce inheritable phenotypes across an enormous range of genetic and environmental variations.

An organism's DNA is buffeted by perturbations that cause significant structural alterations. In humans these lesions occur at a rate of roughly 150,000 per cell per day (Ciccia and Elledge 2010). If left uncorrected, the consequences of this damaged DNA can be severe "These lesions can block genome replication and transcription, and if they are not repaired or are repaired incorrectly, they lead to mutations or wider-scale genome aberrations that threaten cell or organism viability" (Jackson and Bartek 2009, 1071). Cells have developed elaborate systems for detecting the various kinds of lesions and mounting appropriate correcting or mitigating responses. These responses vary from the disruption of mitosis, to splicing and repair of DNA structure, to apoptosis (Branzei and Foiani 2008), and they are specific to the kind and degree of DNA damage. The DNA damage repair system (DDR), like the immune system, endocrine, or thermoregulatory systems, is a highly sensitive goal-directed system. Its goal is maintenance of the structural integrity of DNA, and *that* goal explains its specific activities: "The DNA damage response (DDR) is a signal transduction pathway that senses DNA damage and replication stress and sets in motion a choreographed response to protect the cell and ameliorate the threat to the organism" (Ciccia and Elledge 2010, 180).

DNA sequences, it turns out, are not inherently stable. Instead, their stability is actively maintained by a goal-directed repair system (DDR). DDR implements the specific activities it does precisely because those activities are conducive to the system's goal of maintaining the structure of DNA.

> The stability of gene structure thus appears not as a starting point but as an endproduct – as the result of a highly orchestrated dynamic process requiring the participation of a large number of enzymes organized into complex metabolic networks that regulate and ensure both the stability of the DNA molecule and its fidelity in replications. (Keller 2000, 31)

The constancy of the genome that is evidently crucial to the high fidelity of inheritance is not a primitive feature of the structure of DNA, rather it is the result of the dedicated workings of a highly sensitive, adaptive goal-directed system. An organism's goals explain the constancy of DNA.

Organismal goal-directedness has further implications for the high fidelity of inheritance. In the heyday of gene-centered evolutionary biology it was generally considered that genes severally code for discrete characters: 'gene for …' talk is predicated on this notion. Nowadays it is thought that genes operate in suites of interactive, adaptive gene regulatory networks. Gene regulatory networks manifest an adaptive robustness. Wagner (2011) and Pfennig *et al.* (2010) demonstrate the enormous capacity of gene networks to maintain their characteristic outputs – heritable phenotypes – across an astonishing range of environmental perturbations and mutations. Gene regulatory networks produce their characteristic output by actively compensating for genetic and environmental uncertainties. It appears, then, that the intergenerational constancy of phenotype that is essential for adaptive evolution is underwritten in some significant measure by the adaptive, goal-directed nature of gene regulatory systems.

Reactive genomes

As the conception of the gene as an individual unit of phenotypic control recedes, it is progressively being replaced by a conception of the genome as a highly integrated, adaptive, corporate entity, the 'reactive genome': "[T]he transition that concerns us has involved genomes rather than genes being treated as real, and systems of interacting macromolecules rather than sets of discrete particles becoming the assumed underlying objects of research" (Barnes and Dupré 2008, 8.)

Genomes are complex, adaptive systems. The genome, conceived in this way, is not so much a repository of information for building an organism. It is an open system that exploits all the various resources available to the organism in development. These resources are genetic, epigenetic, cellular, extracellular, and environmental. The production of phenotypes is not the exclusive or privileged province of any of these developmental resources. Rather, all such influences contribute to the production of an organism, under the reactive guidance of the genome as a whole. Genomes respond to and integrate all these cues.

> At the very least, new perceptions of the genome require us to rework our understanding of the relation between genes, genomes and genetics … it has turned our understanding of the basic role of the genome on its head, transforming it from an executive suite of directorial instructions to an

> exquisitely sensitive and reactive system that enables cells to regulate gene
> expression in response to their immediate environment. (Keller in press, 3)

The reactive, adaptive dynamics of genomes is the hallmark of a goal-directed, purposive system (Walsh in press c). We cannot understand the way that genes, cells, whole organisms, or environments contribute to the development of an organism unless we understand how genomes react to these influences. That is to say, an understanding of genome function must be predicated upon the recognition that they are goal-directed systems.

Conclusion

The Etiological Theory of Function appeared for the best part of four decades to be a singular achievement of naturalistic philosophy of science – a problem for biology solved by philosophy. The etiological theory was born of the conviction that biology cannot take its place among the mature natural sciences if it deals in an occult metaphysics of goals and purposes, and explanations that advert to them. This animadversion to teleology is inherited from the inception of the modern sciences during the Scientific Revolution. The physics of Descartes, Galileo, and Newton demonstrated convincingly that no appeal to goals or purposes is required to explain the phenomena of the physical world. The status of biology as a science in good standing seemed to require that biology followed suit and divest itself of teleology. In this context, the Etiological Theory of Function, as first articulated by Ruse and developed over the course of 30 years, appeared to offer biologists just what they needed. It preserves the apparently teleological tenor of a distinctive class of evolutionary explanations, while relieving them of the teleological commitments – biologists can have their cake and eat it too.

To be sure there is a class of genuine biological explanations to which the etiological theory does (arguably) apply. These are adaptive explanations; explanations of the presence, prevalence, or origin of a beneficial trait type in a population. But for the most part, the alleged virtues of the etiological theory are illusory. There are uses of functional concepts in biology that escape the net of the etiological theory. More to the point, there are genuine teleological explanations in biology whose content is lost or distorted when rendered by the schema provided by the etiological theory. The Etiological Theory of Function does not provide a suitably sanitized, univocal translation scheme for teleological talk in biology.

The failure of the etiological theory to provide an adequate reductive account of teleological explanations in biology, however, should not cause panic among biologists and philosophers of a naturalistic bent. Unlike physics (perhaps), biology does not need to be purged of teleology. Indeed to proscribe teleology would be to deny biologists an essential part of their explanatory arsenal. The biology of the twenty-first century reveals that teleological explanation, and its attendant metaphysics of goals and purposes, are an indispensable part of understanding the process of evolution. The goal-directed purposiveness of organisms is an integral component of adaptive evolution. We cannot understand the dynamics of evolution unless we understand how organisms' pursuit of purposes yields evolutionary novelties, underwrites the fidelity of inheritance, and secures the success of development. The stricture against biological teleology has, in my view, done more harm than good; it has led to an impoverished, etiolated conception of adaptive evolution.

The positive case in favor of biological teleology is based upon the discoveries of empirical science. The more we find out about evolution – especially about the previously neglected role of organisms in directing and regulating that process – the more we see an indispensable role for unreduced teleological explanations. The case against teleology, in contrast, is based upon the precepts of sixteenth-century metaphysics. The Scientific Revolution may have bequeathed to us a de-teleologized world-view, but it also left us something much more important, the understanding that metaphysics is beholden to empirical discovery. In a dispute between sixteenth-century metaphysics and twenty-first-century empirical science, it is obvious with whom contemporary naturalists should cast their lot.

REFERENCES

Allen C., M. Bekoff, and G. Lauder (eds.) (1998) *Nature's Purposes: Analyses of Function and Design in Biology*. Cambridge, MA: MIT Press.

Amundson, R. and G. Lauder (1994) "Function without Purpose: The Uses of Causal Role Function in Evolutionary Biology." *Biology and Philosophy* 9: 443–69.

Ayala, F. (1970) "Teleological Explanations in Evolutionary Biology." *Philosophy of Science* 37: 1–15.

Barnes, B. and J. Dupré (2008) *Genomes and What to Make of Them*. University of Chicago Press.

Bedau, M. (1991) "Can Biological Teleology be Naturalized?" *Journal of Philosophy* 88: 647–55.

(1992) "Where's the Good in Teleology?" *Philosophy and Phenomenological Research* 52: 781–805.

Bigelow, J. and R. Pargetter (1987) "Functions." *Journal of Philosophy* 76: 261–301.

Branzei, D. and M. Foiani (2008) "Regulation of DNA Repair throughout the Cell Cycle." *Nature Reviews (Molecular Cell Biology)* 9: 297–308.

Buller, D. (ed.) (1999) *Function, Selection and Design*. Albany, NY: SUNY Press.

Ciccia, A. and S. J. Elledge (2010) "The DNA Damage Response: Making It Safe to Play with Knives." *Molecular Cell* 40: 179–204.

Cummins, R. (1975) "Functional Analysis." *Journal of Philosophy* 72: 741–65.

Godfrey-Smith, P. (1993) "Functions: Consensus without Unity." *Pacific Philosophical Quarterly* 74: 196–208.

 (1994) "A Modern History Theory of Functions." *Noûs* 28: 344–62.

Griffiths, P. E. (1993) "Functional Analysis and Proper Functions." *British Journal for the Philosophy of Science* 44: 409–22.

Hempel, C. (1965) *Aspects of Scientific Explanation*. New York: Free Press.

Hull, D. (1969) "What Philosophy of Biology Is Not." *Journal of the History of Biology* 2: 241–68.

Jackson, S. P. and J. Bartek (2009) "The DNA-Damage Response in Human Biology and Disease." *Nature* 461: 1071–78.

Jenkins, F. A. Jr. and Goslow, G. E. Jr. 1983. "The Functional Anatomy of the Shoulder of the Savannah Monitor Lizard (*Varanus exanthematicus*)." *Journal of Morphology* 175: 195–216.

Keller, E. F. (2000) *The Century of the Gene*. Cambridge, MA: Harvard University Press.

 (in press) "The Post-Genomic Genome." In S. Richardson and H. Stevens (eds.), *The Post-Genomic Age*. Raleigh, SC: Duke University Press.

Kirschner, M. and J. Gerhart (2005) *The Plausibility of Life: Resolving Darwin's Dilemma*. New Haven, CT: Yale University Press.

Kitano, H. (2004) "Biological Robustness." *Nature Reviews Genetics* 5: 826–37.

Kitcher, P. (1993) "Function and Design." In P. French, T. E. Uehling, and H. K. Wettstein (eds.), *Midwest Studies in Philosophy XVIII*. University of Notre Dame Press, pp. 379–97.

Lewens, T. (2005) *Organisms and Artefacts*. Cambridge, MA: MIT Press.

McCarthy, T. (1977) "On an Aristotelian Model of Scientific Explanation." *Philosophy of Science* 44: 159–66.

Miller, R. (1987) *Fact and Method*. Princeton University Press.

Millikan, R. (1984) *Language, Thought, and Other Biological Categories*. Cambridge, MA: MIT Press.

 (1989) "In Defence of Proper Functions." *Philosophy of Science* 56: 288–302.

Monod, J. (1971) *Chance and Necessity*, trans. Austryn Wainhouse. New York: Knopf.

Nagel, E. (1961) *The Structure of Science*. New York: Harcourt, Brace and World.

 (1977) "Teleology Revisited." *Journal of Philosophy* 76: 261–301.

Neander, K. (1991a) "The Teleological Notion of 'Function'." *Australasian Journal of Philosophy* 69: 454–68.

(1991b) "Functions as Selected Effect: The Conceptual Analysts Defence." *Philosophy of Science* 58: 168–84.

(1995) "Pruning the Tree of Life." *British Journal for the Philosophy of Science* 46: 59–80.

Newman, S. (2009) "Complexity in Organismal Evolution." In C. Hooker (ed.), *Philosophy of Complex Systems*. Handbook of Philosophy of Science 10. New York: Elsevier, pp. 335–54.

Pfennig, D. W., Matthew A. Wund, Emilie C. Snell-Rood, Tami Cruickshank, Carl D. Schlichting, and Armin P. Moczek (2010) "Phenotypic Plasticity's Impacts on Diversification and Speciation." *Trends in Ecology and Evolution* 25: 459–67.

Ruben, D.-H. (1990) *Explaining Explanation*. London: Routledge.

Ruse, M. (1971) "Functional Statements in Biology." *Philosophy of Science* 38: 87–95.

(1981) *Is Science Sexist? And Other Problems in Biological Science*. London, ON: University of Western Ontario Press.

(2003) *Darwin and Design: Does Evolution Have a Purpose?* Cambridge, MA: Harvard University Press.

Salmon, W. (1984) *Scientific Explanation and the Causal Structure of the World*. Princeton University Press.

Schlichting, Carl D. and Armin P. Moczek (2010) "Phenotypic Plasticity's Impacts on Diversification and Speciation." *Trends in Ecology and Evolution* 25: 459–67.

Strevens, M. (2009) *Depth*. Cambridge, MA: Harvard University Press.

Wagner, A. (2011) *The Origin of Evolutionary Innovations: A Theory of Transformative Change in Living Systems*. Oxford University Press.

(2012) "The Role of Robustness in Phenotypic Adaptation and Innovation." *Proceedings of the Royal Society B: Biological Sciences* 279: 1249–58.

Walsh, D. M. (1996) "Fitness and Function." *British Journal for the Philosophy of Science* 47: 553–74.

(2007) Teleology. In M. Ruse (ed.), *The Oxford Handbook of Philosophy of Biology*. Oxford University Press, pp. 113–38.

(2012) "Mechanism and Purpose: A Case for Natural Teleology." *Studies in the History and Philosophy of Biology and the Biomedical Sciences* 43: 173–81.

(2013) "The Negotiated Organism: Inheritance, Development and the Method of Difference." *Biological Journal of the Linnean Society*. doi 1111/blj.12811

(in press a) "Mechanism, Emergence, and Miscibility: The Autonomy of Evo-Devo." *Erkenntnis*.

(in press b) "Chance Caught on the Wing: Metaphysical Commitment or Methodological Artefact?" *Synthese*.

Walsh, D. M. and A. Ariew (1996) "A Taxonomy of Functions." *Canadian Journal of Philosophy* 26: 493–514.

West-Eberhard, M. J. (2003) *Developmental Plasticity and Evolution*. Oxford University Press.

(2005) "Developmental Plasticity and the Origin of Species Differences." *Proceedings of the National Academy of Sciences* 102 Suppl. 1: 6543–49.

Woodward, J. (2002) "What is a Mechanism? A Counterfactual Account." *Philosophy of Science* 69: S366–S377.

(2003) *Making Things Happen*. Oxford University Press.

Wright, L. (1973) "Functions. *Philosophical Review* 82: 139–68.

How physics fakes design

Alex Rosenberg

I owe my start in the philosophy of biology to reading David Hull's and Michael Ruse's pathbreaking first books (Ruse 1973) about this subject. From the start I appreciated Ruse's willingness to take on a clear position ("Tell us what you really think, Michael"). It was an example I have tried to follow, all the way down to the present chapter, yet another defense of reductionism.

Physicalist reductionism

Physicalism is the thesis that the physical facts fix all the facts. Reductionism is the thesis that as well as fixing them, the physical facts explain all the facts in some suitably non-erotetic sense of explanation. The qualification 'non-erotetic' expresses the fact that reductionism is a metaphysical, not an epistemic thesis. Everyone, including reductionists, admits that little of biology can as yet actually be reduced to physics or even needs eventually to be reduced to physics in order to be certified as well grounded.

Reductionists accept these two epistemic claims. They concur in the view advanced for example by Elliott Sober (1984) that there are generalizations we would miss, were we to insist that biological knowledge be built up from physical science. Reductionists treat this truth as a reflection on our cognitive powers. It does not qualify their metaphysical claim that biological facts are a species of physical facts.

Most philosophers of biology are physicalists; few are reductionists. They accept that biological facts, states, processes, events, and property-instantiations supervene on physical facts. They deny that they are identical to them. This too is a metaphysical thesis. Not an epistemic one. Philip Kitcher (1984) expressed this view clearly enough when he observed that the irreducibility of evolutionary biology to molecular biology was not a reflection of limits on our cognitive powers. The most powerful arguments for physicalist antireduction turn on the relationship between the theory of natural selection and physical theory. Time and time again, antireductionists have invoked evolutionary facts explainable by natural selection and not explainable by physical law as the fundamental barrier to metaphysical reduction. It is evolutionary facts and regularities we would miss were we to adopt the point of view of the physicist.

Physicalism needs to show that it is only through the operation of the laws of physics that adaptation can emerge. Otherwise, there is space for the possibility that the physical facts have not fixed all the biological facts, that some adaptations might be the result of nonphysical processes. Physicalism needs to show that the operation of physical law is necessary for the emergence of adaptations. Reductionism needs to show how the process of natural selection is in fact the result of the operation of physical law alone. That is, it needs to show that physical law is sufficient for the emergence of adaptation by natural selection. In this chapter I attempt to accomplish both things: to show that physics is necessary and sufficient for all adaptations, and that the only way they can emerge consistently with physics is by natural selection. This will go at least some way towards vindicating physicalist reductionism, a thesis Michael Ruse still held at least as late as 1989.

Physicalist reductionism needs an explanation that starts with zero adaptation and builds up all the rest of the amazing adaptations of biology from the ground up by physics alone. We can't even leave room for "stupid design," let alone "intelligent design," to creep in. If physicalist reductionism needs a first slight adaptation it must surrender the claim that the physical facts (none of which is an adaptation) fix all the other facts. As a matter of fact, as we'll see, Darwin's theory faces the same requirement.

This is a very stringent demand. It goes far beyond the requirement that biology be compatible with physics, not disagree with it. Logical compatibility with physics is something that science requires from biology. Scientists have long demanded consistency with well-established physics as a requirement on all other theories in science. In fact, the nineteenth-century critics of Darwin's theory were eager to adopt the standard of consistency with physics as a way to blow the theory out of the water. One of these opponents was Lord Kelvin, of second law fame. Soon after the publication of *On the Origin of Species*, Kelvin argued that the Darwinian theory of natural selection had to be false. Darwin estimated that at least 300 million years had been required for natural selection to provide the level of adaptation we see around us. (He was off by three orders of magnitude.) But Kelvin thought he could prove that the Sun was no more than 20 million years old. Given the amount of energy it generated and the best theory of heat production available at the time, Kelvin's theory, the upper limit on the Sun's age was 40 million years. So, there could not have been enough time for natural selection to provide adaptations by the process of natural selection. Thus Kelvin refutes Darwin.

Of course Kelvin didn't have the slightest idea what the real source of the Sun's energy is. It was only after World War II that Hans Bethe won the Nobel Prize in physics for figuring out that the Sun is a very long-lasting hydrogen-fusion-driven explosion. But in 1870 Kelvin's objection had to be taken pretty seriously. Explanations of adaptation must be compatible with the correct physical theory, and in 1870 Kelvin's was the best-informed guess about the physics of solar energy. Darwin himself owned up to being very worried about this problem, since he accepted the constraint of consistency with physics as a requirement on any theory of adaptation.

Although Darwinian biologists have not noticed, their theory needs more than mere consistency with physics. It's not enough to show that Darwinian theory gets it mostly right, apart from a few exceptions, a couple of adaptations produced by non-Darwinian processes at work in biology. To close down the wiggle-room, Darwinism needs to show that the only way adaptations can ever happen – even the most trivial and earliest of adaptation – is by natural selection working on zero adaptation. It needs to demonstrate that, given the constraints of physics, adaptation could have no other source than natural selection. We will sketch how such an argument goes in two stages. First we'll show that natural selection doesn't need any prior adaptation at all to get started: beginning with zero adaptations, it can produce all the rest by physical processes alone. Physics is sufficient for adaptation by natural selection. We need only the second law of thermodynamics to do this. Then with the same starting point, the second law, we can show that the process Darwin discovered is, necessarily, the only way adaptations can emerge, persist, and be enhanced in a world where the physical facts fix all the facts. That will close down the wiggle-room for any alternative source of adaptation in the universe. It will also accomplish the reduction of natural selection to physics.

Showing that physics suffices for adaptation by natural selection

The second law of thermodynamics tells us that, with very high probability, entropy, the disorder of things, increases over time.

But the biological realm seems to show the opposite of second-law disorder. It reflects persistent orderliness – start out with some mud and seeds, end up with a garden of beautiful flowers. The ever-increasing adaptation of plants' and animals' traits to local environments looks like the long-term *increase* in order and decrease in entropy. So we have to square the emergence and persistence of adaptation with the second law's requirement that entropy increases.

You could be excused for thinking that if adaptation is orderly and increases in it are decreases in entropy, then evolution must be impossible. This line of reasoning makes a slight mistake about entropy and magnifies it into a major mistake about evolution. The second law requires that evolution produce a *net* increase in entropy. Increases in order or its persistence are permitted. But they must be paid for by more increases in disorder elsewhere. Any process of emergence, persistence, or enhancement of adaptation must be accompanied by increases in disorder that are almost always greater than the increases in order. The 'almost' is added because, as we have seen, increases in entropy are just very, very probable, not absolutely invariable. It won't be difficult to show that Darwin's explanation of adaptation, and only Darwin's explanation, can do this.

Natural selection requires three processes: reproduction, variation, and inheritance, It doesn't really care how any of these three things get done, just so long as each one goes on for long enough to get some adaptations. Reproduction doesn't have to be sexual, or even asexual, or even easily recognized by us to be reproduction. Any kind of replication is enough. In chemistry, replication occurs whenever a molecule's own chemical structure causes the chemical synthesis of another molecule with the same structure – when it makes copies of itself or helps something else make copies of it. This can and does happen in several different ways, in the test-tube and in nature. The one most directly relevant for evolution on Earth is called template matching – the method DNA uses to make copies of itself.

First step: when atoms bounce around, some bind to one another strongly or weakly, depending on the kind of attraction there is between them – their chemical bond. When the bond is strong, the results are stable molecules. These molecules can only be broken up by forces that are stronger than the bonds. Such forces require more energy than is stored by the stable bond between the atoms in the molecule. Breaking down a stable molecule takes more energy than keeping it together.

Second step: occasionally, these relatively stable molecules can be templates for copies of themselves. Their atoms attract other atoms to themselves and to each other so that the attracted atoms bond together to make another molecule with the same structure. The process is familiar in crystal growth. Start out with a cube of eight atoms in a solution of other atoms of the same kind. They attract another four on each side, and suddenly the molecule is a three-dimensional cross. As it attracts more and more, the crystal grows from a small cube into a large one. The crystal grows in a solution through "thermodynamic noise": the increasingly uneven and disorderly distribution of atoms just randomly bouncing around in the solution mandated by the second law. The atoms already in the crystal latch on to ones in crystal in the only orientation chemically possible, making the nice shape we can see when they get big enough.

A crystal molecule doesn't just have to grow bigger and bigger. Instead the molecule can set up chemical forces that make two or more other unattached atoms, which are just bouncing around, bond with one another, making new *copies* of the original crystal. Instead of getting bigger it makes more copies of itself.

The process could involve more steps than just simple one-step replication. It could involve intermediate steps between copies. Think of a cookie that is stale enough to be used as a mold to make a cookie-cutter that takes the stale cookie's shape. It's a template – to make a new cookie. Make the new cookie, then throw away the cookie-cutter, let the new cookie go stale, and use it to make a new cookie-cutter. Make lots of copies of the same cookie using one-use cookie-cutters. You get the idea. (In a way this is just how DNA is copied in the cell, using each cookie-cutter once, except the cookie-cutter is not thrown away. It uses it for something else.)

Physical chemists and organic chemists are discovering more and more about how such complicated structures arise among molecules. They are applying that knowledge in nanotechnology – the engineering of individual molecules. Pretty soon they'll be able to induce molecules to build any geometrical shape they choose. Often the molecule of choice in nanotechnology experiments is DNA. As usual there are people like the late Michael Crichton and other commentators on nanotechnology who gain attention, warning us that we will soon be overrun by self-replicating nano-robots. The reason they claim to be worried is the role of "self-assembly" in building nano-machines.

Chemists building these structures molecule by molecule is remarkable in itself. What is truly amazing is that the structures assemble themselves. In fact this is the only way nanotechnology works. There are very few molecules chemists can manipulate one at a time, putting one next to another, and then gluing them together. All they can do is set up the right soup of molecules milling around randomly ("thermodynamic noise") and then wait while the desired structures are built by processes that emerge from the operation of the laws of physics. Just by changing the flexibility of small molecules of DNA in the starting soup at the bottom of the test-tube, and changing their concentrations, chemists can produce many different three-dimensional objects, including tetrahedrons, dodecahedrons, and Bucky balls – soccer-ball shapes built out of DNA molecules. Of course, what we can do in the lab, unguided nature can do better, given world enough and time.

Replication by template matching is even easier than self-assembly. And it works particularly well under conditions that the second law of thermodynamics encourages: the larger the number of molecules, and the more randomly the molecules of different kinds are distributed, the better. These conditions increase the chances that each of the different atoms needed by a template will sooner or later bounce into it to help make a copy. In fact, "works well" is an understatement for how completely template-replication exploits the second law.

Let's assume that the mixture of atoms bouncing around in a test-tube or in nature is very disorderly, and getting more so all the time, as the second law requires. As the disorderly distribution of atoms increases, the chances of different atoms landing on the template increase too. Most of the time, an atom bouncing into a template of other atoms is too big or too small or too strongly charged to make a copy-molecule that survives. Even if the new atom bonds to the others, the whole copy may break apart due to differences in size or charge or whatever, sending its constituent atoms off to drift around some more, increasing entropy of course. In most cases, in the lab and out of it, this disorderly outcome of instability in copying is the rule, not the exception. The exception is of course a successful duplicated molecule.

Now let's add some variation to the replication. In effect we are introducing mutation in template-copying. Variation is even easier than replication to get going at the level of molecules. It's imposed on molecules during the process of replication by some obvious chemical facts working together with the second law of thermodynamics.

One look at the columns of the Periodic Table of the Elements is enough to show how disorder makes chemically similar but slightly different molecules.

In the table, fluorine is just above chlorine in the same column. They are in the same column because they react with exactly the same elements to make stuff. Chlorine and sodium atoms bond together and made table salt; that means fluorine and sodium atoms will too (the resulting molecule is a tooth-decay preventer). The reason fluorine and chlorine atoms combine with sodium equally well is that they have the same arrangements of electrons that do the bonding. All that means is that if a chlorine and a fluorine molecule are both bouncing around and bump into the same template, they may both bond the same way with other atoms on the template to make similar, but slightly different molecules. A template with chlorine molecules in it could easily produce a copy-molecule that differs only in having fluorine or two where chlorine would normally go. *Voilà* – variation.

When chemical reactions happen billions of times in small regions of space and time, even a small percentage of exact copies quickly come to number in the millions. As does the percentage of slightly varied copies, with one or two atoms of different elements in place of the original atoms. Most of the time the outcome of this is process is wastage – a molecule that doesn't replicate itself or falls apart, just as the second law requires. But sometimes – very rarely – variation produces a molecule that is slightly better at replicating, or one that is just a little more stable.

Now we have replication and variation. What about fitness differences, the last of the three requirements for evolution by natural selection? Fitness is easiest to understand at the level of molecules bouncing around in a world controlled by the second law. Molecules that form from atoms are stable for more or less time. Some break apart right after forming, as a result of strong atomic forces like charge. Some break apart because their bonding is too weak to withstand the force of other atoms that bounce into them or even just pass by. Some "fragile" molecules will remain intact for a while. They just happen by chance to avoid bouncing up against other molecules, ones with stronger charges that pull atoms away from their neighbors. Here again the second law rears its head: as molecules bounce around, any amount of order, structure, pattern almost always gives way to disorder, to entropy. Hardly any molecule is stable for extremely long periods, with the exception of the bonded carbon atoms in a diamond crystal.

There are differences in stability among molecules owing to the variations that inexact replication permits. Differences in stability have an impact on the replication of different types of molecules. A template-molecule produces copies just by random interaction with atoms that bounce into it or pass close enough to be attracted. The longer the original templating molecule stays in one piece – that is, the more stable it is – the more copies it can make. Most of its copies will be just as stable as the original template-molecule since they will be exact duplicates. They will serve as templates for more copies, and so on, multiplying copies of the original.

Of course, just as there are differences in the stability of different molecules, there are differences in their rates of replication. The number of copies of their templates that can be made, and their stability, will depend on their "environments": on the temperature, the local electric and magnetic fields, the concentration of other atoms and molecules around them. Consider two molecules that differ from each other along the two dimensions of stability and ease of replication. The first remains intact on average for twice as long as the second; the second templates twice as many copies per unit of time as the first. Over the same period, they will produce exactly the same number of copies. What will the long-run proportions of molecules of the two types be? It will be one to one. As far as producing copies is concerned, the two different molecules will have equal *reproductive fitness*. And of course, if their differences in stability and replicability don't perfectly balance out, then after a time there are going to be more copies of one type of molecule than of the other.

Molecules randomly bouncing around a region of space, and bonding to form larger molecules, will eventually, randomly, result in a few stable and replicating structures. Their structures will vary, and the variations will have effects on the molecules' stability and replication rates. These differences, in the ability of molecules to stay intact (stability) and to allow for copies of themselves to form (replicability), will change the proportions of molecules in any region of space. If chemical and physical conditions in that region remain unchanged for long enough, the ratios of the different types of replicating molecules will eventually settle down to a fixed proportion. At that point all the remaining replicating molecules in the region will be equally fit to survive – whether owing to their stability or replicability or varying combinations of both, they and/or copies of them persist. In other words, a purely physical process has produced molecular adaptation: the appearance, persistence, and enhancement of molecules with chemical and physical properties that enable them to persist and/or replicate or both. Then, at some point, the chemical environment changes, slightly or greatly: temperatures rise or fall, new and different molecules diffuse through the region, the magnetic field strengthens or weakens. The process of adaptational evolution starts again, thermodynamically filtering for new stable, replicating molecules adapted to the new conditions.

As this process goes on, two other phenomena become inevitable: the size and complexity of the replicating molecules will increase. Eventually there will start to be molecules that enhance one another's stability and/or replication through their chemical relations to one another. There are no limits to the repetition of this process, making bigger and more complicated and more diverse molecules. If conditions are favorable, the result will be really big assemblies of stable and/or replicating molecules, for instance RNA and eventually DNA sequences and strings of amino acids – i.e. genes and proteins.

The rest is history, that is, natural history. The process we have described begins with zero adaptations and produces the first adaptation by dumb luck, sheer random chance that the second law makes possible. It just had to have happened this way if the physical facts fix all the facts, including the facts of adaptation.

Molecular biologists don't yet know all the details of this natural history, or even many of them. Some have been known for a long time. It was in the early 1950s that two scientists – Stanley Miller and Harold Urey – showed how easy it is to get proteins, sugar, lipids, and the building blocks of DNA from simple ingredients available early in the Earth's history. All they did was run an electric current through some water, methane, ammonia, and carbon. Chemists have been designing similar experiments ever since, getting more and more of the building blocks of terrestrial life. Biologists have discovered that the simplest and oldest of organisms on the planet – the archaebacteria – probably first emerged at least 2.8 billion years ago, and still survive in volcanoes at the bottom of the sea. It is there, in such volcanoes at the bottom of the ocean under the highest temperatures and greatest pressure, that one finds chemical reactions spewing out boiling lava and producing the largest quantities of entropy on the planet. This is just what the second law requires to drive thermodynamic noise, and through it to find stable and replicating molecules in a world of random mixing.

How like the evolution of recognizably biological things – genes, cells, replicating organisms, is the molecular process we have described? Well, recognizably biological evolution has three distinctive features.

First, natural selection usually finds quick and dirty solutions to immediate and pressing environmental challenges. More often than not, these solutions get locked in. Then, when new problems arise, the solutions to old problems constrain and channel the random search for adaptations that deal with newer problems. The results are jury-rigged solutions permanently entrenched everywhere in nature.

A second feature of biological evolution is the emergence of complexity and diversity in the structure and behavior of organisms. Different environments select among variations in different adaptations, producing diversity even among related lineages of organisms. The longer evolution proceeds, the greater the opportunities for more complicated solutions to continuing design problems.

A third feature of biological evolution is the appearance of cooperative adaptations and competing adaptations. The cooperative ones are sometimes packaged together in the same organism sometimes separated in two quite distinct ones living symbiotically (like us and the *E. coli* bacteria in our guts). But natural selection also, and more frequently, produces competition between genes, individuals, lineages, populations, and species. There is always that chance that "arms races" will break out between evolving lineages of traits, in which a new random variation in one trait gets selected for exploiting the other traits it has been cooperating with, or a new variation randomly appears in a trait that enables it to suddenly break up a competitive stalemate it was locked into.

Each of these three features is found in the nano-evolution of the molecules. And each persists as molecular assemblies grow in size, stability, complexity, and diversity to produce organic life.

Lock-in of quick and dirty solutions: only a few molecular shapes will ever allow for their own self-assembly and copying (by templating or otherwise). These shapes begin the lock-in that second-law processes will have to work with in building the more stable, more replicable, bigger, and more complicated molecules.

Increasing diversity and complexity: thermodynamic noise constantly makes more and more different environments – different temperatures, different pH, different concentrations of chemicals, different amounts of water or carbon dioxide or nitrogen, or more complicated acids and bases, magnetic fields, and radiation. As a result, there will be a corresponding selection for more and more different molecules. However, they will still be variations on the themes locked in by the earliest stages of molecular evolution.

Cooperation and competition: some of these molecules will even start to work together, just by luck having structures that enhance one another's stability or replicability. Others will explode or dissolve or change their structures when they combine with one another. They will produce new environments that will select for new chemical combinations, ones that cooperate, ones that compete. And so on up the ladder of complexity and diversity that produces assemblies of molecules so big they become recognizable as genes, viruses, organelles, cells, tissues, organs, organisms ... our pets, our parasites, our prey, and our predators ... and us.

To summarize, molecules bouncing against one another inevitably follow a scenario dictated by the second law. Purely physical and chemical processes in that scenario are *all that is needed* for the emergence, persistence, and enhancement of adaptation through natural selection at the molecular level. Where and when molecules of some minimal size emerge, there will be some natural selection for chemical structures that confer more stability and replicability than other chemical structures confer. These chemical properties are *adaptations*: they have functions for the molecules that exhibit them. They enable the molecules to survive longer than less stable ones in the same chemical bath, and to make more copies of themselves than other ones in the same bath. As a result of molecular natural selection, these molecules are better adapted to their molecular environments than others are.

This is an important outcome for physicalist reductionism. It faces the demand of showing how to get the merest sliver of an adaptation from zero adaptation by purely physical, chemical, thermodynamic processes. More of the same processes can build on that sliver of an adaptation to get more adaptations and eventually more robust adaptations. But it can't cheat. It can't just assume the existence of that first sliver. The second law makes the first sliver possible. Variation and selection can take it from there. Now we have to show that this is the *only way* adaptations – molecular or otherwise – can emerge: by exploiting the second law operating on zero adaptations to begin with.

Showing how physics makes natural selection the only way adaptations can arise

The second law makes the merest sliver of an initial adaptation just barely possible. But it makes no guarantees. For all we know it might happen only once every 13.7 billion years in an entire universe. If the first adaptation survives long enough, the second law allows for improvements, but only if they are rare, energetically costly, and just dumb luck.

But physicalist reductionism needs more from physics than just the possibility of adaptation. Physics won't fix all the facts, unless the second law's way of getting adaptations is the only way to get them. Physicalist reductionism needs to show that blind variation and environmental filtration is the sole route through which life could have emerged in any universe governed by the second law. We have to understand why, in a universe made only by physics, the process that Darwin discovered is the only game in town.

There are only three things we need to keep in mind to do this. First, and as already noted, if we are out to explain how any adaptation at all could ever happen, we can't help ourselves to some prior adaptation, no matter how slight, puny, or insignificant. Second, the second law requires that the emergence and persistence of orderliness of any kind be energetically expensive. An explanation of adaptation's emergence and persistence is going to have to show that the process is wasteful. The more wasteful the better, as far as the second law is concerned.

The very first adaptation has to be a matter of nothing more than dumb luck. It had to be the random result of a very, very mindless process, like a monkey hitting the keys of a typewriter randomly and producing a word. Given world enough and time, the bouncing around of things just fortuitously produces adaptations. The second law insists that initial adaptations – no matter how slight, small, or brief – can't happen very often. And the same has to go for subsequent adaptations that build on these. They will have to be rare and random, as well. So, any theory that explains adaptation in general will have to have this feature: that the first adaptation was a fluke, the luck of the draw, just an accident, simply the result of the law of averages. The inevitability of a first, slightest adaptation is ruled out by the second law: to begin with, the second law says that nothing is inevitable, even the heat-death of the universe. More important, the appearance of the merest sliver of an adaptation is an increase in order, and so at most improbable.

None of this will be a surprise to Darwinian theory of course. That's just what the theory says will happen: variations are random, they are almost always small, in fact they are almost always molecular; the first ones come out of nowhere, just as a result of shuffling the cards; mostly they are maladaptive, and only rarely are they adaptive.

The second law also requires that the process through which later adaptations emerge out of earlier ones be energetically expensive and wasteful: expensive because building more and more order has to cost more and more disorder; wasteful because the early steps – the original adaptations on which later ones are built – will be locked in so that less energetically costly ways of building the later adaptations are not available. Every explanation of adaptation will have to share this feature too. It will have to harness a wasteful process to create order.

Now, one has only to examine natural selection as it has unrolled on Earth to see how expensive and wasteful it is. The combination of blind variation and environmental filtration is great at increasing entropy. In fact, the right way to look at the emergence of adaptation on Earth is to recognize that it is the most wasteful, most energetically expensive, most entropy-producing process that occurs on the planet. The evolution and maintenance of adaptations by natural selection wins the prize for greatest efficiency in carrying out the second law's mandate to create disorder. Forget design, evolution is a mess. This is a fact about natural selection insufficiently realized and not widely enough publicized in biology.

Examples are obvious. A female leopard frog will lay up to 6,000 eggs at a time – each carrying exactly half of all the order required for an almost perfect duplicate offspring. Yet out of that 6,000 the frog will produce only on average 2 surviving offspring. Some fish are even more inefficient, laying millions of eggs at one time just to make two more fish. Compared with that a dandelion is miserly with seeds. It will spread only a thousand seeds and produce, on average, one copy of itself. But the human male competes well with profligate fishes. He ejaculates millions of sperm, full of information about how to build a person, almost all capable of fertilizing an egg, and yet 999,999 sperm out of 1,000,000 fail to do it. A high proportion of most organisms go through an entire life cycle, building and maintaining order and then leaving no offspring at all. Everyone has a lot of maiden aunts and bachelor uncles. Insofar as Darwinian processes make reproduction central to adaptation, they rely on a wonderfully wasteful process. It's hard to think of a better way to waste energy than to produce lots of energetically expensive copies of something and then to destroy all of them except for the minimum number of copies that you need to do it all over again.

How about heredity? Another amazingly entropy-increasing process! Molecular biologists know that DNA copy fidelity is very high and therefore low in information loss, compared with, say, RNA copy fidelity. But think of the costs of so much high fidelity. In every cell there is a vast and complex apparatus whose sole function is to ensure copy fidelity: it cuts out hairpin-turns when the very sticky DNA sticks to itself, it proofreads all the copies, removes mutations, breaks up molecules that could cause mutations, etc. In *Homo sapiens* at least *16* enzymes – polymerases – have so far been discovered whose sole functions are to repair different kinds of error that thermodynamic processes produce in DNA sequences. The costs of high-fidelity heredity – both to build the equipment that protects and corrects the DNA sequences and to operate it – are very great, just what the second law requires.

Now let's look at this energy waste on an even larger scale. The evolution of adaptation reflects environmental change over the Earth's history. The vast diversification of flora and fauna is also the result of differences between local environments in different places on Earth. From long before continental drift and long after global warming, environments on Earth change over time and increase entropy as they do so. What is more, once natural selection kicks in, flora and fauna remake their environments in ways that further accelerate entropy increase. When nature started selecting molecules for stability and replicability, it began producing arms races more wasteful than anything the Americans and Soviets could ever have dreamt up. From that time on, there has been a succession of move and countermove in adaptational space made at every level of organization. It has happened within and between every descending lineage of molecules, genes, cells, and organisms. Each line of descent has forever searched for ways to exploit its collaborators and its competitors' adaptations. All that jockeying for position is wasted when one organism, family, lineage, species is made extinct by the actions of another organism, family, lineage, etc. This of course is what Darwin called the struggle for survival.

Add sexual reproduction to the struggle for survival, and it's impossible to avoid the conclusion that Darwinian selection must be nature's favorite way of obeying the second law. Natural selection invests energy in the cumulative packaging of coadapted traits in individual organisms just in order to break them apart in meiosis – the division of the sex cells. Then it extinguishes the traits and their carriers in a persistent spiral of creative destruction. Think about this for a moment: 99 percent of the species that have been extant on this planet are now extinct. That is a lot of order relentlessly turned into entropy!

It's well known that every major change and many minor ones in the environment condemn hitherto fit creatures to death, and their lineages to extinction. As environments change, yesterday's adaptation becomes tomorrow's maladaptation. In fact it looks as if three different cataclysmic events have repeatedly killed off most of the life forms on Earth. The dinosaur extinction 65 million years ago, owing to an asteroid collision on the Yucatan peninsula, is well established. There are no dinosaur bones in any of the younger layers of stone around the world, but there is a layer of iridium – an element found in much higher concentrations in asteroids than on Earth – spread evenly around vast parts of the Earth centered on the Yucatan in the layers of rock 65 million years old. In that layer the iridium is 1,000 times more concentrated than elsewhere above or below it in the Earth's crust. At a stroke, or at least over only a few years, all the vast numbers of dinosaur species, which had been around adapting to their environment beautifully for 225 million years, just disappeared. That's what made the world safe for the little shrew-like mammals from which we are descended. The fossil record reveals a bigger extinction event 500 million years ago on Earth and an even more massive extinction after that, 225 million years ago: the Permian–Triassic extinction in which three-quarters of all ocean-living genera and almost 100 percent of ocean-dwelling species along with 75 percent of land species became extinct. This is order-destroying waste on a world-historical scale.

Long before all this, it was the build-up of oxygen in the oceans and the atmosphere that killed off almost everything on Earth! Build-up of oxygen? How could oxygen be a poison? Remember, yesterday's adaptation can be today's maladaptation. Life started in the oceans with anaerobic bacteria – ones that don't need oxygen. In fact they produce oxygen as a waste product the way we produce carbon dioxide. Just as the plants clean up our mess by converting carbon dioxide into oxygen and water, the environment cleaned up all that oxygen pollution by molecular action, binding the oxygen to iron and other metals. At some point the amount of oxygen waste produced by the anaerobic creatures exceeded the absorption capacity of the environment. As a result they all began to be poisoned by increasing levels of oxygen around them. These bacteria were all almost completely wiped out 2.4 billion years ago, making enough space for the aerobic bacteria, the ones that live on oxygen and produce carbon dioxide as waste. We evolved from these bacteria.

Can any other process produce entropy as fast and on such a scale as natural selection? Just try to think of a better way of wasting energy than this: build a lot of complicated devices out of simpler things and then destroy all of them except the few you need to build more such devices. Leaving everything alone, not building anything, won't increase entropy nearly as fast. Building very stable things, like diamond crystals, will really slow it down, but building adaptations will use up prodigious amounts of energy. Adaptations are complicated devices: they don't fall apart spontaneously; they repair themselves when they break down. They persistently get more complicated and so use even more energy to build and maintain themselves. Any long-term increase in the number of adapted devices without increased energy consumption would make a mockery of the second law. If such devices are ever to appear, besides being rare, they had better not persist and multiply, unless by doing so they inevitably generate more energy wastage than there would have been without them.

This is the very process Darwin envisioned: in Tennyson's words, "nature red in tooth and claw."

Kelvin had the wrong end of the stick when he argued that there has not been enough time for natural selection to produce the adaptation, the order, we see on Earth. What really needs to be explained is the fact that adaptation here on Earth produces so much disorder. The second law does exactly this, by allowing adaptations, but only on the condition that their appearance increases entropy. Any process competing with natural selection as the source of adaptations has to produce adaptations from non-adaptations, and every one of the adaptations it produces will have to be rare, expensive, and wasteful, We'll see that this requirement – that building and maintaining orderliness always has to cost more than it saves – rules out all of natural selection's competitors as the source of adaptation, diversity, and complexity.

Could there be a process that produces adaptations that is less wasteful than the particular way in which Darwinian natural selection unrolled on Earth? Probably. How wasteful any process producing adaptation can be depends on the starting materials, and on how much variation emerges in the adaptations built from them. But every one of these possible processes has to rely on dumb luck to produce the first sliver of an adaptation. In that respect they would still just be more instances of the general process Darwin discovered – blind variation and environmental filtration. A process that explained every later adaptation by more dumb-luck shuffling and filtering of the earlier adaptations would still be Darwinian natural selection. It will be Darwinian natural selection even if the process was so quick and so efficient as to suggest that the deck was stacked. So long as the deck wasn't stacked to produce some prearranged outcome, it's still just blind variation and environmental filtration. Any deck-stacking – a process of adaptational evolution that started with some unexplained adaptation already in the cards – is ruled out by physics.

Only the second law can power adaptational evolution

The attentive reader will have noticed that at the beginning of the last section I said just three things were required to show that, given the laws of physics, the only way adaptations could have emerged is by natural selection. The first two are the requirement that the first and all subsequent adaptations be random and rare events, the second that the process by which adaptations persist and improve be energetically expensive. But the third requirement we need in order to show that natural selection is the only game in town for building adaptations is a rather deeper and less widely noticed feature of the second law.

No matter what brings it about, the process of adaptation is different from the more basic physical and chemical processes in nature. They are all "time symmetrical." Adaptation is not. But the only way a time-asymmetrical process can happen is by harnessing the second law.

A time-symmetrical process is one that is physically reversible. One example of such time-reversible process is well known: any set of ricochets on a billiard table can be reproduced in exactly the opposite order. Here are some more examples: hydrogen and oxygen can combine to produce water, but water can also release hydrogen and oxygen. Even the spreading circular waves made when a drop of liquid falls into a pool can be reversed to move inward and expel the drop upward from the surface. No matter in what order the basic chemical and physical processes go, they can go in the reverse order too.

The second law creates all asymmetrical processes and gives them their *direction in time*. Now, the evolution of adaptations is a thoroughly *asymmetrical* process. Take a time-lapse film of a standard natural-selection experiment. Grow bacteria in a Petri dish. Drop some antibiotic material into the dish. Watch the bacterial slime shrink until a certain point, when it starts growing again as the antibiotic-resistant strains of the bacteria are selected for.

Now try reversing the time-lapse video of the process of bacterial selection for resistance. What you will see just can't happen. You will watch the population of the most resistant strain diminish during the time the antibiotic is present. After a certain point you will see the spread of the bacteria that can't resist the antibiotic, until the drops of the antibiotic leave the Petri dish altogether. But that sequence is impossible. It's the evolution of maladaptation, the emergence, survival, and spread of the less fit.

There is only one physical process available to drive asymmetrical adaptational evolution. That is the entropy increase required by the second law of thermodynamics. Therefore the second law must be the driving force in adaptational evolution. Every process of adaptational evolution – whether it's the one Darwin discovered, or any other, has to be based on second-law entropy increase.

The physical facts – the starting conditions at the big bang, plus the laws of physics – fix all the other facts, including the chemical and biological ones. All the laws of physics except the second law work backwards and forwards. So every one-way process in the universe must be driven by the second law. That includes the expansion of the universe, the build-up of the chemical elements, the agglomeration of matter into dust motes, the appearance of stars, galaxies, solar systems, planets, and all other one-way processes. And that will eventually include, on one or more of these planets, the emergence of things with even the slightest, merest sliver of an adaptation. We can put it even more simply. In a universe of things moving around and interacting on paths that could go in either direction, the only way any *one-way* patterns can emerge is by chance, here and there, when conditions just happen to be uneven enough to give the patterns a start.

These rare *one-way* patterns will eventually just peter out into nothing. Trust the second law. Consider the one-way process that built our solar system and maintains it. It may last for several billion years. But eventually the nice pattern will be destroyed by asteroids or comets or the explosion of the Sun or the merging of the Milky Way with other galaxies, whichever comes first. That's entropy increase in action on a cosmic scale. On the local scale entropy increase will occasionally and randomly result in adaptational evolution. And that is the only way adaptations can emerge in a universe where all the fact are fixed by the physical facts.

Because entropy increase is a one-way street, the second law is also going to prevent any adaptation-building process from retracing its steps and finding a better way to skin the cat. Once a local adaption appears, it can't be taken apart and put together in different, more efficient, less entropy-increasing ways. The only way to do that is to start over independently. Adaptation-building has to produce local equilibria in stability and replication that get locked in, built into the woodwork, like a knot, and have to be worked around in the creation of new adaptations. Natural selection is famous for producing such examples of inferior design, Rube Goldberg complexity, and traits that could only have been "grandfathered in." It's not just the oft-cited example of the blind spot in the mammalian eye resulting from the optic nerve's coming right through the retina. An even more obvious case is the crossing of the trachea and the digestive system at the pharynx. Convenient, only if you like choking to death. We think of the giraffe's neck as an adaptation par excellence. But the nerve that travels from its brain to its larynx has to go all the way down the neck, under the aorta and back up – a 20-foot detour it doesn't need to take.

Any adaptation-creating process has to produce sub-optimal alternatives all the time. It has to do this not just to ensure entropy increase but also to honor the one-way direction the second law insists on. Perhaps our most powerful adaptation is the fact that our brains are very large.

They have enabled us to get to the top of the carnivorous food chain everywhere on Earth. But this is only the result of a piece of atrocious design. The mammalian birth canal is too narrow to allow large-brained babies to pass through. This bit of locked-in bad design meant that the emergence of human intelligence had to await random changes that made the skull small and flexible at birth and thus delayed brain growth till after birth. This is where the large fontanel – the space separating the three parts of the infant's skull – comes in. Now the kid has room to get through the birth canal and has a skull that will immediately afterward expand to allow a big brain to grow inside it. But brain growth after birth introduced another huge problem: the longest period of childhood dependence of any species on the planet. All this maladaptation, corner-cutting, jury-rigging, is required by the second law of any process that produces adaptations.

What does all this come to? The only way a recipe for building adaptations can get along with the second law is by employing it. The only recipe that can do that is the process that Darwin discovered: dumb-luck variation, one-way filtering, and a very expensive copying machine.

Coda: the second law, natural selection, and theism

The process of natural selection is a matter of probabilities in the same way that the spread of gas molecules in the up-stroke of a bicycle pump is a matter of probabilities, and for the same reason: the process of entropy increase is probable, not certain. This will come as no surprise, given the role of the second law in generating evolutionary asymmetries. But even before this role became clear, the nature of evolutionary probabilities was well understood. The theory tells us that fitness differences between organisms will *probably* lead to differences in their reproductive successes. The Darwinian mechanism cannot guarantee the reproductive success of the fitter of two organisms, lineages, or populations. Therefore it doesn't guarantee the evolution of adaptation. If it did claim that fitness differences guaranteed reproductive success, Darwin's theory would be false. Biologists are well aware of several quite rare circumstances under which, just through bad luck, the fittest among competing creatures do not survive at all, let alone have more offspring than the less fit ones. Most of these circumstances fall under the category of "drift."

It is the role of low probability events in evolution that makes it incompatible with Christian theism. This is a view Ruse was not entirely clear on in *Can a Darwinian be a Christian? The Relationship between Science and Religion* (2004). His later writing on a blog post, "Evolution and Christianity: Did We Arrive by Chance?" (2011), is much clearer on this point. In what follows I show how the physical sources of Darwinian natural selection in second-law processes make any reconciliation with theism not just unlikely but impossible.

One feature is arguably indispensable to the theism of all the Abrahamic religions – Judaism, Christianity, and Islam. They all say, "God created man in his own image. In God's image he created him; male and female he created them" (Genesis 27). One thing they agree on is that we exist and have our features owing to God's design, based on an idea – an image, a design – of what he wanted. How literally to take this claim remains a matter of debate among the various sects, creeds, churches, divinity schools, madrassas, and biblical exegetes of the Abrahamic religions. What theism can't give up is the credo that he created us sapient creatures, and that he did so intentionally. Theism cannot accept the notion that we are a side-effect, a by-product, an accident or coincidence, an unintended outcome.

Now, the standard reconciliation of Darwinism and theism goes like this. Everything *else* extant in zoology and botany might well be such a by-product, a side-effect, and an unintended, but foreseen outcome of God's handiwork in creating us. Elephants and *E. coli* are by-products, since they were produced, along with us, by natural selection, and natural selection was the instrument, the means, the technique God used to produce his one intended outcome – *us*. A more ecumenical view might have it that all God's creatures, great and small, are equally intended outcomes, along with us, and all were produced by the device of blind variation and natural selection, just as the omniscient Lord knew they and we would be so produced.

The trouble with this reconciliation is that it does not take Darwin's theory seriously. It just pretends to do so. Darwin's theory tells us that we, and every other creature that roams the Earth or the deep, are at best *improbable* outcomes of natural selection. It is therefore a highly *unreliable* means of making us or anything that looks like us. Indeed it is an unreliable way of making intelligent life of any sort, or for that matter making life of any sort. Any omnipotent deity or even a very powerful agency who decided to employ natural selection to produce us would be disappointed many more than 99 times out of 100 attempts. An omnipotent agency that employed natural selection to make *Homo sapiens* would certainly not be an omniscient one. He would have had to be rather dim. Even weak-minded creatures like us know that blind variation and environmental filtration is not a reliable way of making anything that you have a very definite design for already.

Theism, or at any rate a theism that makes the Abrahamic religions sensible, is just not compatible with Darwin's theory. You can't believe both. But if you are a true believer in theism, you can come close to reconciling them, perhaps close enough to deceive yourself. The trick is to believe that God is powerful enough, smart enough, and present at enough places and times to guide the process of evolution so that it *looks like* random variation and natural selection to anyone not in on the secret. Suppose God employed a method to obtain us – *Homo sapiens* – that was so complex, so subtle, and so difficult to uncover, that the closest we could come to figuring it out was Darwin's theory of blind variation and environmental filtration. You might even go further and suppose that the closest any finite intelligence could come to uncovering God's method was Darwin's theory. You might even suppose that God designed our cognitive capacities so that the only scientific theory of evolution we could ever conceive of would be Darwin's. It would be, as Kant supposed of Newton's theory, the only game in town, not because it is true, but because its falsity is unthinkable by us once we see the evidence for it.

These suppositions would readily explain to a theist why so many smart scientists believe that Darwin's theory is true. For that matter it would also explain why the smartest among them are atheists as well, since they recognize that theism and Darwinism are incompatible. But this stratagem won't reconcile theism and Darwinism.

In fact, one clear implication of this stratagem is that Darwinism is *false*. Darwinian theory tells us that evolutionary outcomes are the result of probabilities – chance reproduction, chance variation, chance environmental events, operating on low population numbers. But the theist thinks that the process God chose was fiercely complicated and absolutely certain to have produced us. As such the theist is committed to Darwin's theory being wrong, false, mistaken, erroneous, incorrect. Darwinian theory is false but just looks correct to smart people trying to figure out what caused adaptation. Saying it's false is no way to reconcile the truth of a theory with theism. Of course, the ways of God may be so far beyond our ken that the smartest and most knowledgeable of us *mistakenly* think adaptation is the result of natural selection. But that wouldn't reconcile theism and Darwinism. At most it shows how to be a theist and to treat Darwin's theory as a useful fiction, one that gets the mechanism of evolution completely wrong.

238   Alex Rosenberg

REFERENCES

Kitcher, Philip (1984) "1953 and All That: A Tale of Two Sciences." *Philosophical Review* 93 (3): 335–73.

Ruse, Michael (1973) *The Philosophy of Biology*. London: Hutchinson.

(1989) "Do Organisms Exist?" *American Zoologist* 29 (3): 1061–66.

(2004) *Can a Darwinian be a Christian? The Relationship between Science and Religion*. Cambridge University Press.

(2011) "Darwinism and Christianity: Did We Arrive by Chance?" www.huffingtonpost.com/michael-ruse/evolution-christianity-chance_b_866301.html (accessed September 30, 2013).

Sober, Elliott (1984) *The Nature of Selection*. Cambridge, MA: MIT Press.

Index

adaptation, 9, 11, 72, 173, 185, 197, 198, 209, 218, 219, 233
  and orchestration, 209
  and the second law of thermodynamics, 220, 225, 232, 234
  conditions of existence, 10
  cooperative and competing, 226
  maladaptation, 230
  molecular, 224, 227
  the physics of, 11, 218
Ariew, 202
Ayala, 3, 4, 13, 203, 208

Bateson, 8, 110, 115, 122, 123
Beatty, 10, 82, 175
Beckner, Morton
  The Biological Way of Thought (1959), 1
Bellah, 46, 47, 49, 52, 56, 57, 58
biometry, 110, 132
Burian, 156, 157

Castle, 7, 87
cognition, 152
consilience, 65, 73, 74, 76, 77, 78, 79, 80, 114, 115
  classificatory vs. theoretical, 78
cultural evolution, 17, 18
  cultural heredity, 17
Cummins, 197, 199, 202, 203

Darwin, 22
  "community selection," 146
  "conditions of existence" as a law, 111
  "the unity of life" as a law, 111
  a central objective, 10
  achitect metaphor, *see* natural selection as an architect
  analogy between artificial and natural selection, 114, 189, *see* natural selection as a breeder
  and embryology, 152
  and Mendelian heredity, 122

consilience of induction, 115
finches, 72
four axioms, 113, 126, 127
Herschel's philosophy of science, 111
Kelvin's objection, 218, 219, 232
Malthusian struggle, 110
moral sense, 145
Newtonian mechanics as the exemplar of a scientific theory, 111
physicalist reductionism, 218
species problem, 82, 83, 127
  nominalist conception, 6, 112, 127
struggle for existence, 127, 132, 230
The Origin, 6, 8, 9, 14, 83, 112, 115, 176, 218
  informal theory, 8, 110, 126, 134
  not a single equation, 109
The Variation of Animals and Plants Under Domestication (1868), 35
theory of heredity
  pangenesis, 116
variation, 162
  developmental explanation, 164
Dennett, 45, 58, 59
design, 9, 10, 11, 173, 175, 194, 199, 217, 218, 226, 229, 234, 235, 236, 237
  and the laws of physics, 11
development
  cell, 153, 210
    apoptosis, 210
    differentiation, 165
    division, 156
    mitosis, 210
    splicing and DNA repair, 210
  computational models
    Boolean networks, 167
  constraints, 162
  cytology, 154, 155
  developmental genetics, 153, 158
  developmental mechanisms, 9, 155, 158, 162, 163, 168
  embryology, 152, 154, 156

development (*cont.*)
  environmental regulation, 155
  epigenetics, 160, 162
  evolvability, 162
  gene regulatory networks, 167, 211
  genotype-phenotype map, 9, 161, 162, 164, 168
  heterochrony, 155
  high fidelity of inheritance, 11, 209, 211, 229
  homology, 152, 162
  hox genes, 156, 158, 166
  modularity, 162
  ontogeny, 155
  origin of variation, 9
  phenotypic accommodation, 209
  plasticity, 162, 208
  reactive genomes, 211
  regulatory genomic system, 166
  structural integrity of the genome, 11, 208
DNA Barcoding, i, 7, 87, 88, 90, 95, 100, 102
Dobzhansky, 13
  "Nothing in biology makes sense except in the light of evolution," i, 2
Duhem, 5, 41
Durkheim, 56, 57, 58, 60

Ereshefsky, 7, 65, 69, 79, 81, 82, 96
evolution
  altruism, 145
  and entropy, 11, 220, 229, 230, 231, 233
  and theology, 29
  chance/purpose tension, 10
  continuous vs. discontinuous, 115, 122, 125, 126, 127, 134
  degenerative, 147
  evolutionary novelties, 5, 158, 167, 208, 209
  fact and theory, i
  fundamentally chancy, 10
  gene flow, 68
  genetic drift, 161, 235
  genetic homeostasis, 68, 69
  Mendelian heredity, 8, 129
  mutations, 36, 37, 71, 72, 127, 163, 166, 168, 209
    random, guided, unguided, 5, 32, 35, 36, 38, 40, 42
    regulatory, 166
  phenotypic evolution, 9, 158, 162, 163, 164, 167
  progressive evolution, 147
  tree of life, 70, 74, 80
  units of evolution, 68
evolutionary theory
  analogy with economics, 8, 137, 140, 141, 142, 143, 146, 148, *see* Fisher
  and atheism, 5, 41, 43

and teleology, 207, *see* teleology
and theism, 11, 41, 235, 237
  incompatible with Darwinism, 236, 237
and thermodynamics, 126
anti-reductionist consensus, 11, 217
as a design theory, 176, *see* design
auxiliary assumptions, 5, 41, 42, 43
causally incomplete, 5, 34, 38, 39
determinism/indeterminism, 33
developmental evolutionary biology ("devo-evo"), 9, 158, 160, 162, 164
evolutionary developmental biology ("evo-devo"), 9, 152, 154, 156, 160
exact vs. inexact science, 126
fidelity of inheritance, 208
hidden variables, 35, 40
  supernatural, 39
historical explanation, 10
logical compatibility with divine intervention, 5, 32
logical compatibility with physics, 218
mechanistic explanation, 158, 162, 164, 165, 167, 168, 206
Modern Synthesis, 9, 155, 158, 164
morphology, 158
origin of variation vs. variants fate, 168
paleontology, 158
phylogeny, 155, 156
population dynamics, 9, 129, 131, 161, 162, 168
probabilistic character, 5, 32, 33, 35, 39, 235
proximate vs. ultimate causes, 169
reductionism, 151, 217, 218
structure, 7, 107
  causal, 8, 162, 168
  mathematical, 8, 128, 140, 144.
    See mathematical models, 76
  equilibrium, 124, 130, 131, 132, 134
thermodynamics, 224
unification, 125, 126, 127, 134
variation vs. selection as explanatory prior, 10

Fisher, 8, 110, 126, 127, 128, 129, 130, 132, 133, 134, 137, 138, 148
  eugenics, 8, 137, 138, 139, 143, 145, 146, 148
  formalisation of Darwinian evolution, 125, 126, 127, 132, 134
  fundamental theorem of natural selection, 132, 133, 137
  Malthusian parameter, 138, 139, 140, 141, 143, 144, 148
  reproductive value, 141, 143, 144
  The Genetical Theory of Natural Selection, 9, 110, 125, 137
function, 10, 173, 193, 194, 195, 196, 197, 198, 199, 200, 202, 204, 208, 209, 212

analysis, 199, 202, 204, 205
and accidental effects, 196
and adaptation, 198
and malfunction, 196
ascription vs. explanation, 202, 204
etiological theory, 10, 195, 196, 198, 203, 212
  problems, 201, 204, 205
evolutionary, 202
functional analysis, 202, *see* Cummins
functional explanation
  as adaptive explanation, 203, 212
  as causal explanation, 197, 201
  as historical explanation, 203
  as teleological explanation, 203, 204,
    see teleology
  deductive-nomological, 197, 200
  normative implications, 196
mechanism, 200, 202, 205
molecular, 227
univocity, 199, 202, 204

Galileo, 48, 51, 109, 110, 212
Galton, 110, 115, 116, 117, 118, 119, 120, 121, 122,
  123, 132
Gayon, 8, 137
genes, 68, 69, 75, 76, 80, 91, 119, 129, 132, 155,
  157, 163, 209, 211, 212, 225, 226, 227
genetic information, 90, 91
Gerhart, 155, 208
Gilbert, 155, 156
Godfrey-Smith, 196, 201
Goudge, Thomas
  The Ascent of Life (1961), 1, 8
Gould, 68, 135, 155
  and Eldredge (1972), 68
Gray, 32, 175, 184, 185
Grene, Marjorie, 1
Griffiths, 79, 196, 201

Haeckel, 153, 154
Haldane, 110
Hall, 135, 155, 161, 162, 164, 167, 168
Hamilton, 25, 97, 100
Hardy, 124, 129, 130, 131
Hempel, 197, 200
Herschel, 111
  *A Preliminary Discourse on the Study of
    Natural Philosophy (1831)*, 111
Hooker, 175, 176, 190, 215
Hull, 1, 13, 45, 65, 66, 67, 68, 69, 73, 75, 81, 151,
  194, 217
human evolution, 152, 185
  altruism, 18, 21, 23
  anatomical traits, 15, 16
  behavioral traits, 16

culture, 16, 19
Denisovans, 15
evolutionary ethics, 3
  ethical behavior, 17
    ability to anticipate the consequences of
      actions, 19
    ability to choose, 21
    ability to make value judgments, 20
    biological roots, 19
  ethical behavior vs. ethical norms, 4, 17, 22
  ethical norms, 22
    analogy with language, 27
    and sociobiology, 23
    naturalistic fallacy, 4, 18, 23, 25
  genome, 14
  Homo erectus, 16
  Homo habilis, 16
  Homo neanderthalensis, 14
  means-ends connection as the fundamental
    intellectual capacity, 19
  sociobiology, 4, 18, 23, 25
Huxley, 115, 125, 155

Johanssen, 115

Keller, 210, 212
Kirschner, 155, 208
Kitano, 208
Kitcher, 5, 11, 45, 70, 81, 199, 217

Laubichler, 8, 9, 151, 161
Lyell, 175, 176, 177

Maienschein, 8, 9, 151
Mayr, 68, 70, 72, 74, 82
Mendel, 8, 110, 115, 118, 122, 123, 127, 130, 131,
  132, 134
Millikan, 196, 200
molecular biology, 158, 160,
  166, 217
Monod, 193, 209

Nagel, 197, 200, 203
natural selection, 83, 161
  and physical law, 218, 227
  and the second law of thermodynamics, 11,
    133, 219, 227, 231
  anti-reductionism, 218, *see* evolutionary
    theory: anti-reductionist consensus
  as a breeder, 175, 176, 181, 188
  as a deductive consequence, 127, 129, 134
  as a mechanism, 115, 194, 200
  as an architect, 10, 175, 176, 178, 179, 181, 184,
    185, 186, 187
  as emulating design, 194, *see* design; teleology

natural selection (*cont.*)
  as the only way adaptations can emerge, 11,
      219, 232, 235
  Darwinian, 232
  evolutionary stable strategy, 21
  expensive and wasteful, 228
  Fisher's demographic approach, 9
  fitness, 33, 132, 133, 134, 139, 140, 143, 235
  group selection, 21
  inclusive fitness, 24, 26
  intergroup selection, 145
  levels of selection, 151, 157
  mathematical models, 33, 128
  primary cause of evolution, 67, 175, 179
  reduction to physics, 219
  replicators and interactors, 68
  requires reproduction, variation, and
      inheritance, 220
  role, 162
  selective advantage, 36
  stabilizing selection, 68
  The Origin, 175
  units of selection, 67
    and the species problem, 67
  variation, 175, 177, 188
    chance, 175, 178, 184, 185, 188, 190
naturalism, 41, 212, 213
Neander, 194, 196, 200

Ockham's razor, 43
organisms
  and complexity, 226
  as goal-directed entities, 197, 203, 208, 209,
      210, 211, 213
  as highly ordered, self-building, negentropic
      systems, 11
  as purposive entities, 11
  homeostasis, 203
  robustness, 211

Paley, 175
Pearson, 110, 115, 119, 120, 122, 123, 127
Physicalism, 217, 218
population genetics, 8, 130, 137, 163
pragmatism, 48

quantum mechanics, 35

Raff, 153, 154, 155, 156, 157, 158
religion
  Abrahamic religions, 3, 5, 32, 45, 236, 237
  "accommodationists," 31, 42, 43
  as false doctrines, 46
  atheism, 45
    "Four Horsemen," 6, 45

compatibility with evolutionary biology, 4
deism, 31, 32, 41
doctrines vs. practices, 46
ethical truth, 59, 60
  as progress, 54
  divine command theory, 53
evidentialism, 42
fictional truth, 50, 51, 56, 60
fideism, 42
religious aptness, 56
religious pluralism, 49
religious truth, 6, 56, 60
  and mythical truth, 57, 58
religiously progressive, 56, 57
  and ethical progress, 57
  purifying progress, 58, 59
scientific vs. mythical truth, 47, 49, 56
secularism, i, 46, 59, 60
The Theology of the Unhidden God, 41
theism, 31
  interventionism, 31, 41, 42
  Young Earth Creationism, 31, 42, 43
Rosenberg, 11, 217
Ruse, 3, 4, 5, 6, 7, 8, 10, 11, 13, 31, 42, 45, 46,
      59, 60, 65, 66, 67, 68, 69, 70, 72, 73,
      74, 77, 78, 79, 80, 83, 111, 130, 151,
      152, 160, 164, 169, 176, 190, 193, 194,
      195, 197, 199, 200, 201, 204, 212, 217,
      218, 235
Monad to Man
    *The Concept of Progress in Evolutionary
        Biology (1996)*, 14
The Darwinian Revolution
    Science Red in Tooth and Claw (1979), 1
The Philosophy of Biology (1973), 1
The Philosophy of Human Evolution (2012), 13

Salmon, 201
Schroedinger, 11
Schutz, 47, 48, 49, 52, 57
Sober, 5, 11, 31, 32, 34, 36, 42, 217
sociobiology, 151
species problem
  as units of evolution, 69, *see* units of
      evolution
  biological species concept, 6, 70, 74
    evolutionary species concept, 74
    genotypic cluster concept, 70
    phylogenetic species concept, 70, 74
  general lineage concept, 77
  historical entities, 7, 66, 71, 80
  individuals, 7, 65, 67, 68, 69
  microbial species concept
    ecological concept, 76
    phylogenetic concept, 76

recombination species concept, 76
natural kinds, 7, 65, 113
nominalist conception, 6, *see* Darwin
path dependency, 71, 72, 80
speciation
    allopatric model, 71, 75
species anti-realism, 7, 66, 80, 83
species can have multiple origins, 70
species realism, 65, 73, 80
Strevens, 207
systematics, 6, 63, 96, 97,
        100, 102, 157

taxonomy, 6, 7, 77, 79, 87, 90, 91, 92, 93, 94, 95,
        96, 97, 100, 101, 102
barcode index number (BIN), 90
biodiversity crises, 97
biological identification
    microgenomic identification, 88
    morphological identification, 88
cladists, 74
DNA metasystematics, 91
informatics approach to biodiversity, 93
information vs. knowledge, 95, 97
molecular taxonomic methods, 93
operational taxonomic units (OTUs), 90
parataxonomy, 99
phenetics, 97
taxonomic practice, 7, 88

teleology, 9, 10, 193, 194, 195, 196, 198, 205, 208,
        212, 213, 215
and backward causation, 203, 207
and causal explanation, 195
ersatz, 10
goal-directedness as a gross behavioral
        property of a system, 206
goals
    as counterfactually robust difference
            makers, 207
    end-states, 206
    intrinsically normative, 206
purpose, 9, 10, 195
reductionism, 195, 213
Thompson, 8, 109

Wagner, 157, 158, 159, 161, 208, 209, 211
Wallace, 110, 115
Walsh, 10, 11, 202, 203, 207, 212
Weinberg, 110, 129, 130, 131
Weismann, 115, 116, 122
Weldon, 110, 115, 122, 123, 127
West-Eberhard, 209
Whewell, 65, 73, 77, 111, 114, 115, 179
    Philosophy of the Inductive Science (1840), 111
Wilson, 23, 25, 26, 79, 87, 93, 97, 102, 151, 153
Wittgenstein, 52
Woodward, 206, 207
Wright, 110, 196

Printed in the United States
By Bookmasters

Printed in the United States
By Bookmasters